The Strange Loops of Translation

The Strange Loops of Translation

Douglas Robinson

BLOOMSBURY ACADEMIC
NEW YORK • LONDON • OXFORD • NEW DELHI • SYDNEY

BLOOMSBURY ACADEMIC
Bloomsbury Publishing Inc
1385 Broadway, New York, NY 10018, USA
50 Bedford Square, London, WC1B 3DP, UK
29 Earlsfort Terrace, Dublin 2, Ireland

BLOOMSBURY, BLOOMSBURY ACADEMIC and the Diana logo are trademarks of Bloomsbury Publishing Plc

First published in the United States of America 2022
This paperback edition published 2023

Copyright © Douglas Robinson, 2022

Cover design: Eleanor Rose

All rights reserved. No part of this publication may be reproduced or transmitted in any form or by any means, electronic or mechanical, including photocopying, recording, or any information storage or retrieval system, without prior permission in writing from the publishers.

Bloomsbury Publishing Inc does not have any control over, or responsibility for, any third-party websites referred to or in this book. All internet addresses given in this book were correct at the time of going to press. The author and publisher regret any inconvenience caused if addresses have changed or sites have ceased to exist, but can accept no responsibility for any such changes.

Whilst every effort has been made to locate copyright holders the publishers would be grateful to hear from any person(s) not here acknowledged.

Library of Congress Cataloging-in-Publication Data
Names: Robinson, Douglas, 1954- author.
Title: The strange loops of translation / Douglas Robinson.
Description: New York : Bloomsbury Academic, 2022. | Includes bibliographical references and index.
Identifiers: LCCN 2021027342 (print) | LCCN 2021027343 (ebook) | ISBN 9781501382420 (hardback) | ISBN 9781501382468 (paperback) | ISBN 9781501382437 (epub) | ISBN 9781501382444 (pdf) | ISBN 9781501382451
Subjects: LCSH: Translating and interpreting.
Classification: LCC P306 .R63 2022 (print) | LCC P306 (ebook) | DDC 418/.02–dc23
LC record available at https://lccn.loc.gov/2021027342
LC ebook record available at https://lccn.loc.gov/2021027343

ISBN:	HB:	978-1-5013-8242-0
	PB:	978-1-5013-8246-8
	ePDF:	978-1-5013-8244-4
	eBook:	978-1-5013-8243-7

Typeset by Integra Software Services Pvt. Ltd.

To find out more about our authors and books visit www.bloomsbury.com and sign up for our newsletters.

Contents

Pandemonial Abbreviations   vii

Introduction   1
I.1   "Paradoxical Level-crossing Feedback Loop"   5
I.2   "Pleasantly Pervasive Paradoxes"   6
I.3   The Strange Loops of Translation   9
    I.3a   First Strange Loop of Translation: Self-reference   11
    I.3b   Second Strange Loop of Translation: The Incoherently Written Source Text   13
    I.3c   Third Strange Loop of Translation: The Passage of Time   16
I.4   The Structure of the Book   21
I.5   Acknowledgments   27

1   The Strange Loops of (Non)Equivalence   31
    1.1   The Campaign Against Word-for-Word Translation   31
    1.2   The Strange Loops of Sense-for-Sense Translation: Jerome   38
    1.3   The Shared Strange Loops of Sense-for-Sense Translation   43
    1.4   Strange Loops of Word-for-Word Translation: Friedrich Schleiermacher   53
    1.5   Conclusion   66

2   The Strange Loops of the Translator-Function   69
    2.1   The Strange Loops of the Translator-Function 1: Myriam Díaz-Diocaretz   71
    2.2   The Strange Loops of the Translator-Function 2: Rosemary Arrojo   84
    2.3   Towards an Author-Function: Derrida, Barthes, Foucault   86
    2.4   The Strange Loops of the Translator-Function 3: Theo Hermans   94

3   The Strange Loops of Translation as (Peri)Performative Identities   101
    3.1   Logical Aporias and the Strange Loops of Periperformative Workarounds: Mauricio Mendonça Cardozo   102

	3.2	The Strange Loops of Translating Heidegger's Untranslatables: Sabina Folnović Jaitner	109
	3.3	The Strange Loops of "Good" and "Bad" (Periperformative) Translatabilities: Natalia S. Avtonomova and Tatevik Gukasyan	116
	3.4	The Strange Loops by which Translation Shapes Collective Subjectivities: Sakai Naoki and Lydia H. Liu	127
4	The Strange Loops of Translating Bodies		137
	4.1	The Strange Loops of Somatic Response: The DRP	139
	4.2	The Strange Loops of Knowledge-Translation as Mouthable Rhythm: Henri Meschonnic	146
	4.3	The Strange Loops of the Translator's Constructivist Agency: Kobus Marais	158

Conclusion: The Strange Loops of Translation as Transgressive Circulations: Johannes Göransson	179
C.1 Hoaxes	179
C.2 Interiority and Identity	183
C.3 Transminoritization	187
C.4 Salutary Failures	189
Notes	192
References	201
Index	215

Pandemonial Abbreviations

AAP	Amittai Aviram Pandemonium
AaBP	Aase Berg Pandemonium
ABP	Antoine Berman Pandemonium
AKP	Aleksis Kivi Pandemonium
AP	Aristotle Pandemonium
ARP	Adrienne Rich Pandemonium
ASP	Allan Schore Pandemonium
ASSP	Aleksandr Sergeyevich Pushkin Pandemonium
BBP	Brian Baer Pandemonium
BMA	Brian Massumi Assemblage
CHP	Carol Hofstadter Pandemonium
CLP	Comparative Literature Pandemonium
CPP	Chaim Perelman Pandemonium
CSP	Christina Schäffner Pandemonium
CSPP	Charles Sanders Peirce Pandemonium
DAP	Dante Alighieri Pandemonium
DDP	Daniel Dennett Pandemonium
D&GP	Deleuze&Guattari Pandemonium
DHP	Douglas Hofstadter Pandemonium
DMCP	Don Mee Choi Pandemonium
DRP	Douglas Robinson Pandemonium
DTP	Daniel Tiffany Pandemonium
EANP	Eugene A. Nida Pandemonium
EAP	Emily Apter Pandemonium
EBP 1	Emile Boisacq Pandemonium
EBP 2	Emile Benveniste Pandemonium
EOP	Eugene Onegin Pandemonium
EP	Editor Pandemonium
ESHP	Eusebius Sophronius Hieronymus Pandemonium
ESP	Eve Sedgwick Pandemonium
FHP	Friedrich Hölderlin Pandemonium

FSaP	Françoise Sagan Phenomenon
FSP	Friedrich Schleiermacher Pandemonium
GSP	George Steiner Pandemonium
GTP	Gideon Toury Pandemonium
HCP	Holden Caulfield Pandemonium
HGP	Holy Ghost Pandemonium
HMP	Henri Meschonnic Pandemonium
HTP	Hothead Translator Pandemonium
JDP	Jacques Derrida Pandemonium
JGHP	Johann Gottfried Herder Pandemonium
JGP	Johannes Göransson Pandemonium
JLP	Jacques Lacan Pandemonium
JMP	Joyelle McSweeney Pandemonium
JVSP	J.V. Snellman Pandemonium
KHP	Kim Hyesoon Pandemonium
KMP	Kobus Marais Pandemonium
KTP	Kizito Tekwa Pandemonium
LCP	Lying Cretan Pandemonium
LHLP	Lydia H. Liu Pandemonium
LHP	Ludvig Holberg Pandemonium
LTP	Literary Translator Pandemonium
LVP	Lawrence Venuti Pandemonium
MCEP	M.C. Escher Pandemonium
MDDP	Myriam Diáz-Diocaretz Pandemonium
MFP	Michel Foucault Pandemonium
MHP	Martin Heidegger Pandemonium
MKP	Mia Kankimäki Pandemonium
MLP	Martin Luther
MMCP	Mauricio Mendonça Cardozo Pandemonium
MMP	Melissa Mesku Pandemonium
MMPP	Maurice Merleau-Ponty Pandemonium
MSP	Mark Schluter Pandemonium
MTCP	Marcus Tullius Cicero Pandemonium
MTP	Maria Tymoczko Pandemonium
NCP	Noam Chomsky Pandemonium
NSAP	Natalia S. Avtonomova Pandemonium

PMP	Patrick Mensah Pandemonium
PP	Plato Pandemonium
QHTP	Quintus Horatius Flaccus Pandemonium
RAP	Rosemary Arrojo Pandemonium
RBP	Roland Barthes Pandemonium
RWEP	Ralph Waldo Emerson Pandemonium
S&TAP	Source&Target Authorial Pandemonium
SAP	Source Author Pandemonium
SFJP	Sabina Folnović Jaitner Pandemonium
SHP	Stefan Heyvaert Pandemonium
SNP	Sakai Naoki Pandemonium
SRP	Source Reader Pandemonium
TAP	Target Author Pandemonium
TGP	Tatevik Gukasyan Pandemonium
THP	Theo Hermans Pandemonium
TMP	Timothy Morton Pandemonium
TRP	Target Reader Pandemonium
UEP	Umberto Eco Pandemonium
VKP	Volter Kilpi Pandemonium
VNP	Vladimir Nabokov Pandemonium
WBP	Walter Benjamin Pandemonium
WSP	William Shakespeare Pandemonium
WWP	Walt Whitman Pandemonium
WWoP	William Wordsworth Pandemonium

Introduction

The notion of a "strange loop" was first theorized in 1979 by Douglas R. Hofstadter in *Gödel, Escher, Bach* (*GEB*) as occurring "whenever, by moving upwards (or downwards) through the levels of some hierarchical system, we unexpectedly find ourselves right back where we started" (1979/1989: 10)—and it caught on, sort of. Political philosophers like Morden (1990) and Grundmann and Mantziaris (1991) found Hofstadter (1979/1989) useful for their thinking. A surprising number of literary scholars have found hermeneutical applications for strange loops, among them Dixon (1989), Bethlehem (1999), Littler (2013), and Williams (2018). Le Moigne and Sibley (1986) applied strange-loops theory to the problems of information management in corporations, Launer (2010) to the convoluted dynamics of the psychotherapeutic session. Not a flood of applications of the main theoretical innovation in this brilliant and popular book—but some.

Nothing, however, in the four decades since that initial publication, on the strange loops of translation.

The reason that is worth mentioning, as if it were some sort of culpable failure of uptake, is that beginning in the mid-1990s Hofstadter, or what we'll be calling the Douglas Hofstadter Pandemonium (they/them pronouns), began both to publish on translation and to move toward a rethinking and reframing of the strange-loops theory in *GEB*. Indeed, even in *GEB* they had recurred several times to the problematics of interlingual (372–3, 379–80) and intersemiotic translation (83), but also, and above all, to translations between and among mathematical systems in Kurt Gödel's engagement with *Principia Mathematica* (209–13, 215, 267–72, 417, 441–6, 533); other explicitly translational analogues in that book included string theory (234–6), biology (485, 518–48), and mechanics (380, 601, 603). But no strange loops of translation.

In *Le Ton Beau de Marot* (Hofstadter 1997), a 600-page book devoted entirely to translation, they dealt in passing with many of the cognitive quirks that they had previously linked to strange loops—especially analogy and indexicality/ self-reference—but without theorizing translation *as* a strange loop. As that translation book was about to come out in the spring of 1997 they began to translate Pushkin's *Onegin* into English; when that translation was released in 1999, the Translator's Preface said nothing about strange loops, but also nothing about the cognitive quirks that had complicated their thinking about translation in *Le Ton Beau de Marot*.

In a preface to the twentieth-anniversary edition of *GEB* published in 1999, they complained about readers missing the book's strangely loopy point (see Hofstadter 2007: xiii–xv for this story); and in response to that 1999 preface, in the spring of 2003 Ken Williford and Uriah Kriegel asked them to write a continuation of their strange-loops theory for a philosophical essay collection they were editing, titled *Self-Representational Approaches to Consciousness*. When that book came out in the spring of 2006, it contained a double-length article by the Hofstadter Pandemonium, who felt that it still wasn't long enough; their further work on the topic resulted in *I Am a Strange Loop* (2007). Again, that work contained passing analogical mobilizations of translation—translating in a broad sense, especially between Gödel's theorem and English (145) and as an analogue of the replicability of patterns (224, 257), but also twice referring specifically to literary translation[1]—but no link between translation and strange loops.

While they were working on the article for Williford and Kriegel (2006), they decided to translate Françoise Sagan's 1966 novel *La Chamade* into English—a labor of love, not a commission from a publisher. Some time between finishing that translation in early 2005 and publishing it in 2009, they wrote their second extended treatise on translation, titled "Translator, Trader"—only a hundred pages long, considerably shorter than *Le Ton Beau de Marot*—which they published in an omnibus volume with the Sagan translation. Their main focus in it was how they had translated Sagan, with extended discussions (culminating in defenses) of some of their choices (the proliferation of Americanisms, most notably, by their own count at a rate of around twenty per page). But they also theorized the "paradoxes" of translation—without, again, building a theoretical bridge from those paradoxes to strange loops.

This strikes us as a recurring but ultimately simple oversight. Our purpose in this book is to remedy that oversight.

A preliminary note on pandemonial pronouns. We will be referring to each diffracted and displaced strangely looped persona mentioned in this book in the plural as nominative *they*, accusative *them*, and possessive *their*, on the dual grounds that (a) in theorizing the (plural) strange-loops constitution of the (singular) "I," Douglas Hofstadter—which as we've noted we'll be calling the Douglas Hofstadter Pandemonium and abbreviating "the DHP"—tacitly borrowed something from their friend Daniel Dennett (1991) on the "pandemonium" self (each of us has inside *lots* of self-demons, each saying "I" but tonalizing it divergently); and (b) there is something intriguingly queered and queering, transgendered and transgendering, about the argument from strange loops (see Robinson 2019). To the extent that strange loops, as the DHP themselves insist, are multiply and recursively (self-)constitutive of consciousness, and thus of the voice that serially says "I," they (b) *trans* identity. Strange loops, translation(s), and transgender(s)—in the (usually clandestine) plural.

We are not, however, suggesting with the phrase "the Douglas Hofstadter Pandemonium" that Hofstadter is somehow *uniquely* pandemonial—viz., full of demons, like hell in *Paradise Lost*, where John Milton coined the term Pandemonium to mean "the place of all the demons." We are suggesting instead that we are all that way. We also refer several times, even in this note, below, to "the Douglas Robinson Pandemonium," or DRP. We will also continue referring to ourselves with the plural "we"—not the editorial we, let alone the royal we, but the pandemonial we. Every hour of our existence, even when we are sleeping, we are generating and partly launching, partly discarding thousands of self-demons saying "I." And by "we" we mean all of us, including y'all. We will be using that useful Southernism throughout, y'all, for the reader-demons of this book. The DRP lived in Mississippi for twenty-one years, and are planning to retire there. Sorry if y'all find that plural "you" irritating.

Though Dennett is the DHP's close friend, and though Dennett's *Consciousness Explained* is cited twice in *I am a Strange Loop*—the first time the DHP telling us that they read the 1991 book "in manuscript form" (219), the second time on 315—they do not explicitly mention the "Pandemonium model" developed by the Daniel Dennett Pandemonium (DDP) there. We only claim that the DHP "tacitly borrowed" that model because it seems like such a deeply congruent way of thinking about the strange-loops constitution of the "I":

> In the Pandemonium model, control is usurped rather than delegated, in a process that is largely undesigned and opportunistic; there are multiple sources for the design "decisions" that yield the final utterance, and no strict division is possible between the marching orders of content flowing from within and the volunteered suggestions for implementation posed by the word-demons.
>
> (Dennett 1991: 241; quoted in Robinson 2001: 155)

Those "multiple sources for the design 'decisions' that yield the final utterance" are what the DDP call the "word-demons" that vie ballistically for primacy in any given utterance; and because the (un)design of utterances, and thus of speech acts, and thus of human social interaction is ballistic in the sense of being *impossible to control after launch*, the DDP's demons throw scare quotes up around "decisions." We humans like to think that we have something to say, something that we have formulated carefully in advance and *decide* to utter in a certain way, but no: the design is ballistic, and *multiply sourced*.

The Pandemonium model says that we are all plural ("we"/"they"), even though we like to idealize ourselves as singular ("I"/"he"/"she"). Not only that: while the DHP note that the title *I Am a Strange Loop* is shorthand for "'I' Is a Strange Loop" (Hofstadter 2007: xv), we would add that their actual theorization of the "I" throughout the book actually makes the title shorthand for "the ostensible singularity of the 'I' is constituted iteratively through a serial multiplicity of strange loops." Viewed in this light, perhaps, the DHP's strange loops might be usefully understood in part as more brilliantly complex articulations of the DDP's demons.

Throughout this book, therefore, all singular pronouns and verb forms channeling those "I"-dealizations will appear in scare quotes. Limiting ourselves otherwise to (unmarked) plural pronouns and verbs for the multiple minds in singular bodies often makes for awkward articulations, but we've decided to risk that awkwardness here, because the benefits for the strange-loops argument strike us as overwhelming. Indeed, so intrinsically pluralized is that argument as the DHP themselves develop it in *I Am a Strange Loop* that the "I"-demons of the Douglas Robinson Pandemonium (DRP) are mostly surprised that the DHP demons didn't embrace plural pronouns themselves, at least in *I Am a Strange Loop*.

In that DRP mix, however, are other "I"-demons that are quick to point out the eagerness with which the DHP in "Translator, Trader" not only articulate but aggrandize the "I," like dogmatic apostles of the Sovereign Ego.

We're sorry to say that those rebel DRP demons extend the same hypocritique to the DRP's own self-aggrandizing speech-act history as well. In truth, it requires of us a massive effort—for which we keep foolishly holding out hope of some day receiving commensurate credit (aka "the moral high ground")—to refer to ourselves exclusively in the first-person plural.

Additional note. See https://xn--wgiaa.ws/3-doug-robinson-strange-loops-of-translation-douglas-hofstadter for an online version of this introduction as published by Melissa Mesku in 👁👁👁 ("many loops") with a toggle feature that allows y'all to select pandemonial or "original" pronouns and personal names.

Thanks to Melissa for not only publishing it there but coming up with the idea of a toggle feature, for readers who find the pandemonial pluralizations weird and annoying. If that's y'all, unfortunately there's no way to toggle in a print book: y'all are stuck here.
End of note.

What strikes us, then, and inspires this book, is that despite the intensification of their thinking about both strange loops and translation throughout the first decade of the new millennium, the DHP persist in what we take to be the obvious oversight in *Le Ton Beau de Marot*: they never raise the possibility of the strange loops of translation. In *I Am a Strange Loop* (2007: 101–2) they had defined a strange loop as "a paradoxical level-crossing feedback loop"; the subtitle of "Translator, Trader" (2009) two years later is "Essays on the Pleasantly Pervasive Paradoxes of Translation." That suggests a potential theoretical convergence—but one, unfortunately, that is never even broached, let alone explored.

I.1 "Paradoxical Level-crossing Feedback Loop"

Logically, as the DHP define it, a strange loop is not just any old paradox but specifically a *cyclical* paradox:

> What I mean by "strange loop" is … not a physical circuit but an abstract loop in which, in the series of stages that constitute the cycling-around, there is a shift from one level of abstraction (or structure) to another, which feels like an upward movement in a hierarchy, and yet somehow the successive "upward" shifts turn out to give rise to a closed circle. That is, despite one's sense of departing ever further from one's origin, one winds up, to one's shock, exactly where one had started out. In short, a strange loop is a paradoxical level-crossing feedback loop.
> (2007: 101–2)

The logic, however, is still not the main point—not what the DHP complained that readers of *GEB* had mostly missed. The main point that they returned in *I Am a Strange Loop* to underscore was that these strange loops *in* human consciousness are constitutive *of* consciousness, and thus of the "I"—that ultimate symbol(izer/-ized) of consciousness.

The first sign of strange loopiness, they suggest, is the presence of indexicals, words pointing to the work being done by the speaker (or thinker) to point to their own current symbols, such as "this" or "here," but also "I" (160); but these words are only indications that symbols are registering symbols, and especially that a higher-level symbol, called "I," is thinking not only about symbols, but about its own thinking about those symbols. What makes this process strangely loopy is the pandemonial proliferation of analogies (148–52) or isomorphisms (243) between lower-level symbolization and higher-level symbolization: whatever an "I" says about the symbols it is/they are manipulating can easily be applied to the "I" saying those things, so that any attempt to idealize the "I" as a god-given glory that is inserted into the human organism at birth, or at conception, and that may even survive the death of the organism into which it was so inserted, is everywhere undermined by its own continuous self-creation.

The second important thing to note there is that the analogical parallels, the isomorphisms, that give the game away are only logical in secondary idealizations. First and foremost they are *audience-effects*. They are situated interactive performativities. And the implication of that realization is that the "I" itself is also a performativity, an audience-effect. It is a perceptual multiplicity (the pandemonial "they") presented and ultimately experienced as a unity (the ontologized "I"). It/they come/s to feel like a "real person" inside the human organism only gradually, as it/they internalize/s how it is/they are perceived by others. The "I," in other words, is a rhetorical construct. The DHP don't identify it as such, or as an audience-effect, presumably because they aren't in the habit of using theatrical metaphors, or of thinking in rhetorical terms; but it's pretty clear that that is the general neighborhood in which their metaphorical imagination is playing. Aristotle would have called it *ēthos*—"character," but also authority, and identity—and argued that it is a joint production of the speaker and their audience.[2]

I.2 "Pleasantly Pervasive Paradoxes"

The subtitle of "Translator, Trader" (2009), y'all will recall, is "Essays on the Pleasantly Pervasive Paradoxes of Translation." The DHP's first paradox is what they call "*The Wrong-Tongue Paradox*" (9), which they gloss with the rhetorical question "How can Dante Alighieri have written a book in English, a language that didn't even exist when he was alive, or William Shakespeare a book of sonnets in Russian, when he knew not a word of that language?" Yes, good questions. But now take it up a notch. What happens when Dante doesn't just write a book in English, but *tells us in English* that they are writing their book in Italian (rather than Latin)? Enter the strange loop. (The DHP come closer to hinting at that possibility in 1997: 445–7—but there too hover just shy of making the connection with strange loops.)

A small difference, y'all say? Maybe. But the paradox the DHP invoke in 2009 is a curiosity. It makes the reader scratch their head and say "huh!" It is odd, but easy to ignore. Most readers of translations never notice it. The DRP's souped-up strangely loopy paradox is in the reader's face. But even that is not the main point to be made about it. Even if no one notices the strange loops of the Dante Alighieri Pandemonium (DAP) saying in English that they're speaking Italian, those loops do cognitive work that "*The Wrong-Tongue Paradox*" doesn't do. Translational self-reference is cognitively productive. Like all strange loops in the DHP's accounts, it is constitutive of the self. And because it is *collectively* self-constitutive, because it constructs the self as a *shared audience-effect*, it is more broadly constitutive of culture.

That culture-constitutive effect is especially clear in another paradox mentioned by the Douglas Hofstadter Pandemonium. They call it "*The Wrong-Place Paradox*":

> When characters in a story speak, the language they use is informal and colloquial, so their lines are inevitably peppered with colorful words and idioms that come from the streets of their land and the history of their people. And any other people has a different history, different traditions, and scads of different colorful ways of expressing itself. Therefore, when idioms of culture B are placed in the mouths of individuals from culture A, the result is incoherent.
>
> (2009: 9–10)

This is another curiosity, of course, one that is well known to translators and translation scholars, but apparently unknown to ordinary target readers: "And yet as readers, we are expected to skip right over that—and the curious thing is that we usually do! The incoherence passes invisibly right through our filters, and we often see nothing wrong at all" (10). Any Literary Translator Pandemonium must inevitably engage this paradox, and somehow work around it; and since the DHP demons have done some literary translating, and are commenting in this 100-page booklet on the 200-page literary translation with which it shares an omnibus volume, they know it quite well. It is no longer invisible to them, and they want to make it visible to their readers as well. A salutary desire!

They are, however, not at all interested in how that paradox twists around into a strange loop that is constitutive of cultures. Why do we think of "culture

A" and "culture B" as discrete entities? Why is French culture separate from English culture? Why does it seem so odd and even deliciously absurd to us in *Monty Python and the Holy Grail* when King Arthur and their knights arrive at a castle in England where everyone not only speaks French but seems to think they are in France? In *Critical Translation Studies* (Robinson 2017c) the DRP demons track the work of Sakai (1997) showing historically how translation practice constructed "Japan" and "the Japanese language" and "Japanese culture" as national entities; what we don't track there is how that constructionist history proceeded through strange loops. We'll be exploring the strange loops of that history in section 3.4.

As we'll see in section 1.1, in fact, the DHP tend to *resolve* these paradoxes by urging the translator to translate sense for sense. We will show there that their resolutions not only fail to resolve the paradoxes, but twist and loop them even more tightly, and even more strangely. But they don't always even recognize the paradoxes in which translation is mired. Consider, for example, this passage:

> The pithy slogan "Translator, traitor" shows very clearly that a translator need *not* be a betrayer or traitor, for it beautifully preserves the key quality that makes the original Italian phrase so memorable—namely, its catchiness, which is due to the fact that the two nouns inside it sound so much alike. There is no important aspect of the phrase *Traduttore, traditore* that is missed by "Translator, traitor," and so this English translation is a checkmate in response to the strong-seeming check tendered by the Italian opponent.
>
> (Hofstadter 2009: 18)

We would argue that it's not actually a checkmate, because a checkmate ends the game, while the elegant translation of *traduttore, traditore* as "translator, traitor" keeps looping back around strangely, potentially without end. By calling the translator a traitor *in* a nontraitorous translation, the target text "translator, traitor" both confirms and disconfirms the Italian slander, and does so cyclically, very much like the liar paradox. The two words of the Italian dictum (source text) say to the English translator (target author), "y'all can't translate us elegantly, because in translating y'all invariably commit treason against us";[3] and the two words of the target text, "translator, traitor," which implicitly *contradict* that accusation ("See, we can translate y'all without treason!"), also explicitly *reiterate* the accusation, returning us back to the beginning: "Ha, see," the source text word-pair retort, "y'all agree with us!" Just as the source text operates simultaneously on two levels, the propositional ("y'all can't translate without treason") and the stylistic (the eloquent consonance of "trad(u/it)tore"), the target text, in attempting to operate equivalently on the same two levels, proves the Italian proposition both wrong and right, wrong stylistically (by mimicking it) and right propositionally (by reiterating it). Every attempt the target text makes to rise above the propositional accusation through eloquent propositional and stylistic equivalence returns it to the starting point; any potential attempt the target text might make to deviate from propositional equivalence would simply prove the source text right.

I.3 The Strange Loops of Translation

So where might we begin to look for the strange loops of translation?

The initial answer is quite easy: if the strange loop of consciousness is generated by analogical parallels between the symbols an "I" is thinking about and the "I" itself thinking about them, the strangeness of that loop is only compounded when the translator inserts their "I" in between a source author and a source reader and deflects or redirects the rhetorical situation from the source culture to a target culture.

The DHP's reflections on "fidelity" in their translation of Sagan into highly Americanized English are rife with examples of this. When Sagan's character Antoine says "et il ne dit rien," for example, Hofstadter bristles at the notion that they should be "faithful" to Sagan by translating that "and he said nothing," as the novel's previous English translator did. "I, by contrast," they note, "stray a bit further from Sagan, and write 'and he bit his tongue'" (2009: 64).

What we find striking there is that the DHP imagine themselves straying further not from Antoine's words, nor from the narrator's report of Antoine's words, nor even from Sagan's creation of the narrator's double-voicing of Antoine, but from Sagan: from the authorial "I." The DHP demons have translated Pushkin's *Onegin* (Hofstadter 1999), but in their preface to that translation they explain that their Russian is minimal, and they *studied* the verse novel closely and *memorized* many stanzas, making it unlikely they've read Mikhail Bakhtin (1929/2002; Emerson 1984 in English) on double-voicing, so that's a bit of a red herring; there are apparently also not enough literary scholars among the DHP demons to distinguish between authorial voice and narratorial voice (cf. 2009: 85). But surely they recognize the analogical shifts that give us both level-jumps and significant parallels between Sagan's "I" and Antoine's "I"? Certainly, they seem to be fervently aware of the analogical jumps/parallels between Sagan's authorial "I" and their own translatorial "I": when they ask, rhetorically, "But do I, a mere translator, have the right to turn up the clarity and vividness knobs?" (64), their response comes very close to theorizing the rhetorical construction of the translatorial "I," but falls short of that strangely loopy goal, and in falling short arguably provides tacit support for what the 2007 DHP repeatedly derogated as the "illusion" or "hallucination" or "sleight of mind" of the stable ontological "I": "Well, the fact is that I'm naturally inclined to turn these knobs up high no matter what I'm writing, because clarity and vividness are, in some sense, my religion. I would be betraying *myself* if I didn't allow myself to be as clear and as vivid as possible when I translate" (64).

But, regarding that "arguably": let's not jump to conclusions here. Let's give the DHP the benefit of the doubt. "Naturally" there is probably a colloquialism that shouldn't be read as implying the DHP's belief that their "I" is a naturally occurring entity with specific stable qualities implanted there by "nature." The author of *Gödel, Escher, Bach* and *I Am a Strange Loop* would certainly not want to be caught objectifying or essentializing (or even singularizing) their own "I"! (It may be "naturalized," in the sense of coming to *feel* natural through frequent repetition and resulting habitualization, but it is never "natural.") And the assertion that "clarity and vividness are, in some sense, my religion" is saved from the opprobrium of objectivism and essentialism not only by the hedge "in some sense" but the very notion that it is a "religion"—something that Hofstadter too would agree is not so much a stable set of ontological truths as it is a set of conventionalized and normativized social practices, beliefs, and values. (Calling clarity and vividness their "religion" doubles up the rhetorical constructedness undergirding "religion": as a social construct and as an analogue of the "I.")

But even the normative invisibilizing of the translatorial "I" is loopily constitutive of the translatorial "I." The more closely we study the actual verbal practicalities of translation, in fact, the stranger the loops become. As we've noted, raising the DHP's "*Wrong-Language Paradox*" to the status of strange loopiness would mean not only tracking the impossible loops as a curiosity, but following the 2007 DHP themselves in tracking the constitutive force such pandemonial strange loops wield in the generation and maintenance of personal and cultural identities. More than that, too, it means envisioning and experimenting with alternative translational strategies for *mobilizing* that constitutive force for new cultural expressions, orientations, behaviors, norms and values, and so on.

I.3a First Strange Loop of Translation: Self-reference

For example, think of the complications that arise whenever the source author uses indexicals: "this," "here," "I," "now," and so on. When Martin Luther writes in their "Sendbrief vom Dolmetschen" (1530/2004)/"Circular Letter on Translation" (Robinson 1997/2014: 88), "Wenn ich nun den Buchstaben nach, aus der Esel Kunst sollt des Engels Wort verdeutschen, müsste ich so sagen: Daniel, du Mann der Begierungen, oder Daniel, du Mann der Lüste," by which of these three logics should we translate that into English?

a. *German-German*: "If I now Germanize the angel's word literally, as these jackasses would insist, should I say: 'Daniel, du Mann der Begierung'?"
b. *German-English*: "If I now Germanize the angel's word literally, after the letters, as these jackasses would insist, should I say: 'Daniel, you man of desires'?"
c. *English-English*: "If I now Anglicize the angel's word literally, as these jackasses would insist, should I say: 'Daniel, you man of desires'?"

Only the a-model there functions logically in terms of the Martin Luther Pandemonium's (MLP's) choices—after all, they translated the Bible into German—but "du Mann der Begierung" is the MLP's (mocking) German translation, which the target reader presumably can't read. Only the c-model functions logically in terms of the choices made by target readers—after all, they read the text in English—but Luther didn't translate into English. There is no logical solution. The "preferred" (normative) translation would follow the b-logic, which of course is completely illogical, because "you man of desires" is not German. The conceptual "workaround" (actually more of a "let's just not think about it") that is tacitly invoked to justify this illogic in normative translation practice is that this is just one of the unresolvable paradoxes of translation, sigh. And, after all, if a scene in an American movie is set in Germany, where Germans are speaking to Germans, and the American actors all speak English with German accents, we accept that they are "actually" (imaginarily) speaking German. So what if it's not "real"? We can live with the paradox. We can pretend. We can suspend disbelief. (This would be more or less the approach championed by the DHP's "solution" to the paradox.)

The MLP's letter in German is a *Sendbrief*, literally a send-letter: one sends it around to people. In English it is a circular letter, a letter that circulates. In the First Strange Loop, what circulates is a layered and problematically distributed attribution of the translated translations in the letter to the MLP's author-function, the MLP's translator-function, and the DRP's translator-function (those Foucauldian concepts are defined and explored by the DRP's author-function in more detail in Chapter 2). The imagined hierarchical upward-chain of inference is:

from step 1
the DRP's bc-translation ("Daniel, you man of desires" [88])
up to step 2
the MLP's sarcastic/mocking a-translation ("Daniel, du
Mann der Begierung")
up to step 3
the angry/ironic claims made explicitly by the MLP's author-function about
the standard or established (literalist) translator-function of their day
("den Buchstaben nach, aus der Esel Kunst"/literally "after the letters, out of
the ass art," or, more loosely, "literally, jackass-style")
up to step 4
the self-justifying claims made implicitly by the MLP's author-function
about their own translator-function, which would watch how German
mothers' mouths move when they speak German and translate accordingly

Because the DRP's translator-function is also assigned the task of rendering that move from 2 to 4 into English, however, the target reader's pleasant passage up those latter steps in the idealized hierarchy is disrupted, because all of a sudden we find the MLP's German author-function describing the German activity of their translator-function as if it had all happened in English: "Now if I wanted to translate this like the jackasses insist, literally ... " (the DRP's actual translation on 88). The DRP's c-translation in step 1, in other words, returns the analogical stair-stepping up the hierarchical chain of inference from the implied higher levels (the MLP doing this or that in steps 2–4) to the DRP's own English-speaking translator-function back in step 1. And if we try hard enough *not* to be disruptive in that strangely loopy way, if the DRP try to preserve the logic of the MLP's author-function by writing, "Now if I wanted to Germanize this like the jackasses insist, literally ... ," they/we still have to decide whether to give the MLP's suggested translation in English (again disrupting the hierarchy) or in German (foreclosing on our translator-function and not serving the target reader); and if the DRP take that latter course, the first step up the hierarchical chain of inference is missing, and the target reader trips and falls.

Of course, there is also the potential disruption caused by the fact that in all three of the DRP translations "Martin Luther" is speaking English—what the DHP call "*The Wrong Tongue Paradox*." This, too, might disrupt target readers' idealized chain of hierarchical inference, if they begin to realize that what the DRP translator-function is leading them to is not the *real* "Martin Luther" but the MLP's English-speaking author-function, a kind of author-function demon-prosthesis as it were. But, as the DHP themselves point out, most target readers never notice this problem. The disruptions that the DRP author-function called the First Strange Loop of Translation are the more flagrant ones.

Then again, considering that the DHP's author-function, which first noticed and theorized strange loops, is also a translator-function and attentive reader of their own translations, and they didn't notice (let alone theorize) these supposedly "more flagrant" disruptions, perhaps the DRP author-function is just imagining their disruptive effect?

I.3b Second Strange Loop of Translation: The Incoherently Written Source Text

The incoherently written source text is, of course, a disturbingly common occurrence in technical translation—something that the demons of the DHP's translator-function have either never tried their hand at or never thought interesting enough to theorize. What is strangely loopy in this case is the marketplace norm: the normatively idealized translator-function must both translate the text equivalently and fix it up, so that the translation gives target readers the impression that the interiority of the source author's "I" has been clearly and faithfully communicated to them. Needless to say, a well-written target text is not equivalent to a badly written source text, and the "clarity and vividness" of the translator-function's idealized "I" is not equivalent to the pandemonial murkiness of the sloppy source author's unidealized "I"! But that's the desideratum.

This Second Strange Loop is akin to the *Drawing Hands* sketch of M.C. Escher, which the DHP reprint and discuss as a classic example of the strange loop (Hofstadter 2007: 102–3). Just as the viewer's response to the Escher drawing is to climb what appears to be a "natural" hierarchy from drawing to hand to arm to artist, but ends up at the point of another pencil and another hand holding it and drawing the first hand, so too is the target reader's response to the "fixed-up" translation to climb from the coherent target text toward what they take to be a coherent source text, and beyond that to the source author's intended meaning, and beyond that to the full serene interiority of the source author's "I"—only to find themselves precisely where the translator started, at the point of *creating* a coherent text out of the incoherent original.

Ideally, of course, the translator-demons are normatively expected to contact the source author and ask a series of clarifying questions: "What'd y'all mean here?" and "Which interpretation of this syntax is correct?" This would engineer the *construction* of a coherent authorial "I" by means of an activist participatory audience-effect, which would in turn engineer the transformation of the pandemonial audience-effect into a (submerged/implicit) translatorial "I" in the target language.

But the writers of incoherent technical texts typically have no coherent public author-function "I" that they can represent normatively in writing. Whether they are verbally incoherent in both their speaking and their writing or only their writing, there is an inarticulacy in or about them that is constitutive of a fractal or fractured "I," a disorganized and disruptive pandemonium whose demons are all pulling in opposite directions. These are most often nameless technicians buried deep in the bowels of some gigantic corporation, often in fact nameless teams of technicians who hate to write and do it on reluctant assignment, through gritted teeth. (Back in the mid-1990s the DRP had a good friend who was a tech writer and had worked in several of these companies before realizing that the cultures there were death for a writer's professional competence and self-respect, and going freelance.) Sometimes the badly written source texts are themselves second- or third-generation translations from Japanese or Chinese, cobbled together too fast and with too little professionalism by those nameless technicians—often, in fact, these days by free online machine translation apps with little or no post-editing. Even if by some fluke there is an identifiable source author, they are typically sequestered behind a translation agency's firewall, lest the freelancer hired to translate their output try to "steal" their business (bypass the agency so as to maximize profits). And if by some miracle the translator does manage to contact the actual living, breathing source author, it is axiomatic that writers who submit incoherent texts for public use are incapable of even recognizing, let alone explaining, let alone resolving, syntactic confusions.

So what are the norm-obedient translator-demons to do? How does the pandemonial but professionally responsible translator produce a pandemonial but coherent authorial-becoming-translatorial "I" that is arguably faithful to the incoherent source text?

They invent it. They rewrite the text coherently, based on their own best guesses.

But surely this is *progress*? Surely, in the DHP's terms, this is *not* a case where "one winds up, to one's shock, exactly where one had started out," and so not a strange loop?

No, because the marketplace requirement is that the translation be *equivalent*. If the translator-demons invented the translation's coherence, then it's not equivalent, and by definition not a translation; but since the translator was hired to produce a translation, then the text their demons create and submit *is* a translation, and by definition also *is* equivalent; and so on, in potentially infinite recursions.

Gideon Toury's (1980: 63–70) ingenious suggestion was that all texts regarded as translations are by default equivalent to their originals—even if there *is* no original, as when a pseudotranslation is "mistakenly" regarded as a translation—because equivalence is the normative assumption that attends the translation-designation. The presumptive translationality of a text makes it equivalent, which makes it a translation of its proclaimed source text—say, the collection of Ossian poems—which is admired greatly through the self-constitutive loop of the translation until a plausible enough hoax accusation creates a big enough backlash audience-effect to make the presumed source text vanish like smoke on the wind.

Thus no matter which way we try to straighten the loop out, to iron out its strange bugs, we keep ending up exactly where we started out—even though that place makes us extraordinarily uncomfortable.

Again, though, the cognitively and affectively productive question is: what does our discomfort *generate*? If the strange loops of layered analogical perception generate the "illusion" or the "myth" or the "hallucination" of the authorial "I," what does our discomfort with the strange loops of fix-it-up technical translation constitute? What is the constitutive effect of those strange loops on the translatorial "I," which has been professionally formed to submerge itself in (and thereby loopily work to constitute) the putative authorial "I"? Does the translatorial "I" begin wanting *more*? More money, more recognition, more joy? Does the newly awakened (but still vestigially suppressed) translatorial "I" begin to assert its "rights" more aggressively, more egotistically, saying things like "I would be betraying *myself* if I didn't allow myself to be as clear and as vivid as possible when I translate," and "were I told that I had to adopt the principle of such rigid 'faithfulness' to the author, then I would just give up translating, for it wouldn't allow me to use my own mind" (Hofstadter 2009: 64–5)?

I.3c Third Strange Loop of Translation: The Passage of Time

Finally, should the translator of, say, Homer's *Odyssey* or the Bible archaize the target language, because the source text is archaic? Or should they modernize it, because (we're told) the source text was not *written* to be archaic, and was originally heard by source listeners as ordinary contemporary speech? The marketplace norm in this case vacillates, but tends to lean more toward modernization—or rather, toward a *cautious* kind of modernization that is also just elevated enough to create the illusion (audience-effect) of being old and venerable. Nothing slangy; nothing impenetrably ancient; nothing in the target reader's face: just enough hints in both directions to allow the target reader to feel comfortable projecting some kind of idealized normative coherence onto the text.

Even those modernizing loops are strange, in fact. Their strangeness has a lot to do with the notion, perpetuated among others by the Walter Benjamin Pandemonium (1923/1972; Robinson 2023 in English), that translations age but originals do not. Because this is a temporal loop, we might well apply to it what Umberto Eco in "The Myth of Superman" (1979) theorizes as a temporal "iterative scheme": just as Superman and the other characters in the strip undergo experiences that in the real world would age them but they never age, always starting back from scratch in the next strip or the next comic book or movie, so too does each modernizing retranslation seek to restart the original back at "scratch," at some imaginary starting point that is to be understood as its "true" "origin." The rhetorical construction of this restart is that it corrects for the "temporal distortions" caused by the aging of past translations—but then it too ages, and it too must be replaced by a new retranslation that restarts the loop. This retranslational iterative scheme seeks, through an audience-effect that is highly vulnerable to the passage of time, to *stabilize* the source text, to fix it in time, so that it is perceived as not aging—but it is perceived that way only *through* the rhetorical ministrations of each new retranslation. Because each new retranslation presents itself as capturing the true originary and stable essence of the source text, what the target readers who ratify that presentation experience, even for a fleeting moment, is the fortuitous convergence of source and target texts at a starting point in time that is surreptitiously (ideally) out of time. For those target readers who do not ratify the image of serendipitous convergence, the new retranslation is just another retranslation—and of course the numbers of these latter readers for each new retranslation tend to swell over time.

The strange loops of archaizing translation work along similar lines, when what the translator is approximating is not some primordial time-out-of-time origin at which the source text was modern but its current archaic flavor. One fairly widespread—even locally normative—pandemonial target-reader orientation is to say that the *Odyssey* and the Bible are not only ancient texts but valuable specifically for their hoary antiquity, and therefore should be archaized in translation as well. What is then being rhetorically stabilized is not an imaginary past origin-moment at which the source text was "modern" but an imaginary present noble-rust moment (see Schlegel 1791/1962: 85–7; Robinson 1997/2014: 214 in English) at which the source text is always respectably "old." A classic example would be Rudolf Borchardt's 1923(/1967) *Dantes Commedia Deutsch*, with its radically archaizing aim of reimagining German history had Dante written originally in German. According to George Steiner (1975/1998: 357), Borchardt's archaizing project "is not antiquarian pastiche, but an active, even violent intrusion on the seemingly unalterable fabric of the past"—but of course that violence was inflicted not on some ontological "fabric of the past" but on an imaginary past as an audience-effect constructed in the present. And to the extent that German readers in the mid-1920s ratified Borchardt's archaic German as "German Dante," they were participating in the stabilization not of some past German Dante but a present-day archaic German Dante—Dante in contemporary (1920s) German culture as old *and* in German. And like the "true" "originary" "essence" of the source text as rhetorically stabilized by modernizing retranslations, that present archaic German Dante is/was subject to temporal changes in German reading practices over time; anyone seeking to restabilize that present archaic German Dante would need to retranslate it, and somehow convince target readers that *now*, finally, the archaic German Dante had finally converged and merged with archaic Italian Dante.

Where the strange loop of archaizing textuality becomes especially loopy, however, is in texts deliberately written to be archaic. In this sort of source text the archaizing strategy becomes an implicit self-reference: "I am archaizing this text," we can imagine the Source Author Pandemonium (SAP) implicating by and through their writing strategy; or, more explicitly, "I am writing this text in Year X in Culture A, and imitating the style of Year/Decade/Century Y, in order to convince you that this is a temporally accurate reproduction of the source text." Authorial archaism, in other words, is a complex kind of culturally situated and loopily performative self-reference: "We are performing our 'I' as simultaneously existing in both *this* time and place in which we now stand and *that earlier* time and place in which our characters and/or imaginary source readers stand, and hope you will join us in that dually situated temporal performativity." The complexity of that performativity is especially salient because typically the culture of the earlier time and place is performed (by both the source author and the source reader: a dual audience-effect) as *historically* constitutive for the writing "I" in the present time and place.

And all that, of course, applies only to the source side. What happens when the translatorial Target Author Pandemonium (TAP) insert themselves into the middle of that performativity, mediately, and partially hijack it into another invented (target-cultural) "past" as imaginary prologue for target readers in another "present" (*this*) time and place (*now*)? What authorial, narratorial, and translatorial "I's" are constituted through this hijacking, and what do the demons of the Target Reader Pandemonium (TRP) naturalize as the "true" or "authentic" "self" or "core" of the source-becoming-target text, and how does that naturalization work? Or is it possible that the TRP naturalize a split "I"? The Pandemonium model posits not just the existence of fragments or fractals of the various "I's," but the ballistic *launching* of those fractal "I's" as demons, all jostling for position and control of articulation. In translation the question becomes whether the rhetorical audience-effect cohesion of those demons into a single coherent "I" would depend on the coherence of the authorial SAP's "I," or of the translatorial TAP's "I." Or would any relative coherence be a *peri*performative "hallucination" or "illusion" built collectively out of the three-way rhetorical interaction, the circulation of evaluative audience-effects among SAP, TAP, and TRP? (We will be mobilizing the periperformative as theorized by Eve Sedgwick [2003] in Chapter 3.)

This Third Strange Loop is obviously a strangely loopy escalation of what the DHP call "*The Wrong-Place Paradox*," in which, as the DHP put it, because the source text is saturated in the history of the source culture and the target culture lacks that history, "when idioms of culture B are placed in the mouths of individuals from culture A, the result is incoherent." We would demur on that "incoherence," which is really a problem in the Second Strange Loop, not this one. (We suspect that what they mean by "incoherent" is not so much confusing and messy but anachronistic, anatopistic, illogical [mismatching] and therefore putatively incoherent from the perspective of eternity.) Escalating the DHP's talk of "paradox" into a strange loop entails an exploration of the ways in which "the result is [not] incoherent" but merely recursively complex; adding the strange loopiness of deliberate archaism further complicates that exploration.

A few years ago the DRP faced something like this problem in translating Finland's greatest novel, Aleksis Kivi's *Seitsemän veljestä* (1870), as *The Brothers Seven* (Robinson 2017a; see also Robinson 2017b).[4] The Aleksis Kivi Pandemonium (AKP) had written the novel in the 1860s about their parents' generation, using a stylized southern Finnish dialect from the 1840s to flesh out a paleo lifestyle that by their day was almost defunct. What to do in English?

Archaic texts are notoriously difficult to read, and lectorial TRPs typically expect the translatorial TAP to make things easier for them; indeed, some translation scholars call the lexical simplification of the source text—"the language is usually flatter, less structured, less ambiguous, less specific to a given text, more habitual, and so on" (Pym 2010: 79, summarizing Toury 1995: 268–73)—a *universal* of translation. This makes archaizing translation strategies counternormative, at least in a generalized "marketplace preference" sort of way.

The interesting tension in the market's equivalence norm, however, is that it requires the translator to capture both the *semantic content* and the *style* of the source text. The commonsensical (low-end) response to that norm is that it is impossible to achieve both; hence the standard retreat from the full implications of the norm into dumbed-down sense-for-sense equivalence, which is entirely compatible with lexical simplification and stylistic flattening. The greatest literary translators have historically sought to achieve both— apparently mutually exclusive—goals, but typically with a greater willingness to fudge semantic equivalence where necessary in order to achieve something like stylistic equivalence (or perhaps just stylistic brilliance).[5]

That latter (or "higher") approach would seem to require that the translator archaize, and thus create a text that TRPs—beginning with acquisitions editors— will typically find excessively difficult. The AKP's Finnish is hilarious, but also quite difficult for the "ordinary" or "uneducated" or "untrained" demons of the Finnish Source Reader Pandemonium (SRP) to read, and the previous two (low-end) translations, from 1929 by Alex Matson and 1991 by Richard Impola, do make it simpler. In that way, arguably, they do the English TRP a great service. The novel is much easier for TRPs to understand in their English than it is for SRPs in the AKP's Finnish.

Unfortunately, however, Matson and Impola also purge the novel of all humor, and indeed of all other literary qualities as well. Maybe that, then, is not such a great service? If the "universal" of translation is to reduce a brilliantly funny but difficult novel to tedious, turgid, pedestrian simplicity—flatter and more habitualized than the original, with rhythms and other prosodics deadened, with alliterations and other euphonic effects dulled through a tin ear, with the delectable paleo vocabulary dumbed down through a deluge of high-frequency lexical items—maybe translation is "universally" not such a great thing? Maybe then the Comparative Literature Pandemonia who have long warned us against translations have always been right? Maybe the only answer for anyone who loves literature is to learn the source language and read the work in the original?

Maybe not. What the DRP did with the AKP's novel was to translate it into a playful reconstruction of Shakespearean English—from the early seventeenth century, much earlier than the AKP's 1840s Finnish rural dialect. Why? Partly because the AKP were saturated in the William Shakespeare Pandemonium (WSP) in Swedish translation, had memorized several whole plays in Swedish, and often translated quotes from the WSP's plays into their Finnish texts; and because their archaic Finnish was always playful, always designed to make the reader laugh out loud, and in every way strongly indebted to Shakespearean humor. (Also to Cervantean humor, but we weren't translating them into Spanish. Had we been, we would certainly have used Cervantean Spanish.)

Partly, too, our strategy was based on the assumption that English readers would be at least slightly familiar with the WSP, and would enjoy a playful Shakespearean English more than some obscure 1840s Scots dialect, say.

Mainly, though, our strategy was motivated by the fact that we, too, love the WSP, much as the AKP did. Gearing up to tackle the considerable challenge of this translation, we assumed—perhaps projecting this onto the AKP—that they and we both felt the "same" thrill in the brilliant edge of the WSP's writing, the restlessness of their intelligence, the explosive audacity of their humor. Prompted by that "shared" love, we blended the mutual self-constitution we experienced between "our" WSP and ourselves with the mutual self-constitution between "our" AKP's WSP and "our" AKP, all of us creating and organizing the others' "I's" and "we's" in playful interaction, or active interplay. We read the AKP reading the WSP, we read the WSP reading the AKP, they read us reading both, and gradually the strange loopiness of all that anachronism constituted the

"I" of the AKP who shadowed us as the DRP was translating through the word-demons of a partly invented Shakespearean English.

Or, all right, we did it all. It all happened in our head (one head, many demons). Like the translator who invents a target coherence for an incoherent source text, we invented a playfully archaic style for our translation that radically *exceeds* the original—that gives the impression of archaizing equivalence only as a periperformative audience-effect. Progress? Not really a translation? Not, therefore, a strange loop? No, because again, it's a translation, and therefore equivalent. Back to square one.

And again, the big question is not only how that strange loop shaped (reconstituted/collectivized) our translatorial "I," but what effect it might have on target readers' "I's," and thus on the target culture.

I.4 The Structure of the Book

To the casual reader it may seem, from our discussions of Hofstadter on translation—especially "Translator, Trader"—that the book is set up as a hit job on Hofstadter's thought. Not so. The DRP's demons are great admirers of the DHP's cognitivist theorizing—even when they/we disagree with it. Those demons also find the DHP's remarks on translation in *Gödel, Escher, Bach* and *I Am a Strange Loop* cognitively exciting, and their remarks on literary translation in *Le Ton Beau de Marot* witty and artistically inventive. There are, to be sure, DRP demons that read "Translator, Trader" as somewhat unfortunate; but in the aggregate the DRP demons are fans of the DHP's work, and it is precisely as fans that they/we would like to make a contribution to its aggregate success, especially by building the bridges the DHP have so far neglected to build between their theorizing of translation and their theorizing of strange loops.

To that end we have set the book up as a series of dialogues between Hofstadter on strange loops and a parade of thinkers primarily about translation but in the background more generally about literature, language, and subjectivity. The idea is for each dialogue to be mutually enriching: for Hofstadter's theories to serve as a complicating inspiration for the thinkers whom we put into dialogue with them, but for those thinkers to complicate and enrich Hofstadter's theories as well. Hofstadter has a Ph.D. in physics, and knows more about mathematical theory than anyone else mentioned in this book; they have channeled their STEM background into stunningly original and transformative theories of crowd-sourced subjectivity-formation. Everyone else in the book knows a lot more about language, literature, and translation. Mutual enrichment: it's not a competition.

Chapter 1, "The Strange Loops of (Non)Equivalence," sets up a series of increasingly radical dialogues on what is still arguably the dominant issue in translation studies: equivalence. Section 1.1, "The Campaign Against Word-for-Word Translation," tracks the DHP's rather primitive attack on translational literalism as *"words ... trigger[ing] words without thoughts intervening,"* and on imaginary demands that translators render word for word as "denying what makes us human"; much of the rest of the chapter gives the lie to the DHP's dire prediction that "any translation based on such a philosophy will turn out arid and wooden at best, and absurd and incomprehensible at worst" (37). The interesting thing about the Hofstadter Pandemonium's defense of sense-for-sense translation is not just that it misses the strange loops in which it is mired, but that it seems to the DHP demons to *solve* the paradoxes that threaten it—and, by solving them, prevents them from turning into full-fledged strange loops. (Not that the DHP ever mentions the threat of translational strange loops; that account of "preventing them from turning into strange loops" is a speculative reading of the DHP's puzzling total foreclosure on the strange loops of translation. Perhaps the DHP demons had an inkling of those strange loops, but took that inkling as a problem in need of a solution, and grasped at sense-for-sense translation as the traditional panacea?)

Jerome, of course, was the coiner of the term "sense-for-sense translation," and in section 1.2, "The Strange Loops of Sense-for-Sense Translation," we explore the implications of the ESHP's "exception": the demons of the Eusebius Sophronius Hieronymus Pandemonium translate sense for sense, they tell us, *"except of course* in the case of Holy Scripture, where even the syntax contains a mystery." Oops. And if that weren't bad enough, the ESHP demons play both ends against the middle, turning Scripture and non-scripture, mystery and non-mystery, and sense-for-sense and word-for-word translation into Möbius strips, aka strange loops.

There is also the odd fact that the DHP, who worked so hard in *I Am a Strange Loop* to undermine the notion of the stable sovereign "I," seems two years later, in "Translator, Trader," not only to crown their own self-appointed translator-function as precisely the kind of sovereign "I" that the earlier book had lambasted as a pernicious myth, but to set the translatorial "I" *in opposition to* the authorial "I": "It's not as if I were rewriting the plot of Sagan's novel, or changing her characters' characters; I am just being myself" (2009: 65). In addition to the attack on the "I" as a "myth" and an "illusion" and a "hallucination," *I Am a Strange Loop* features the DHP's lengthy and powerfully moving speculation about the *sharing* of personalities, specifically in the context of their marriage to Carol, who died suddenly at the age of forty-two—a speculation that we explore as a different potential model for sense-for-sense translation in section 1.3, "The Shared Strange Loops of Sense-for-Sense Translation." The traditional model of the "true" translator is that they have no "I": they let the source author speak through them, like sound waves transmitted through water or air, or a current conducted through a nerve or a wire. The DHP's "just being myself" model in "Translator, Trader" would seem to mark the self-assertive opposite pole to that traditional notion. But what about the DHP's own "shared personalities" model as an account of sense-for-sense translation? Wouldn't that chart an interesting middle ground between the poles, with source author and translator forming a composite authorial/translatorial "I"? If so, would that entail a paradisal or mystical unity that might avoid strange loops altogether? Or would it, too, be an audience-effect constituted rhetorically through strange loops?

The DHP's notion that literal translation is mindless, like Google Translate—that the literal translator simply replaces each source-textual word in sequence with its "obvious" target-textual equivalent *without thinking*, mechanically—is wildly at odds with the actual history of literal translation. The first such historical counterexample, launched in section 1.4, "The Strange Loops of Word-for-Word Translation," is Friedrich Schleiermacher, whose impassioned plea for a literalist German National Translation Project in their 1813 address before the Royal Academy of the Sciences in Berlin, "On the Different Methods of Translating," is anything but mindless. So intensely do the Friedrich Schleiermacher Pandemonium (FSP) demons support their literalism (later rebranded "foreignism"), in fact, by giving their sense-for-sense opponents numerous near-hysterical *pieces* of their mind, that we offer as a marital objective correlative of their model—an alternative to the "paradisal" marriage explored in section 1.3—the marriage of two people who hate each other but stay together for decades out of strangely looped caution and self-protection.

Chapter 2, "The Strange Loops of the Translator-Function," tackles the extension offered by Myriam Díaz-Diocaretz (1985) to Michel Foucault's theory of the author-function, and the three later critical engagements with that extension, all fortuitously published within a few months of each other in 1997: by Rosemary Arrojo, Theo Hermans, and the author of this book. The interesting question raised in the chapter is whether the translator-function should be regarded as an ideal model that an actual translator or Target Author Pandemonium (TAP) should attempt to emulate, as it seems to be for Díaz-Diocaretz; a branding tool by which the TAP can attempt to enhance their visibility, as it is for Arrojo; or a normative straitjacket that can be used to police possible deviations in TAPs, as it is for Hermans. In section 2.3 we also take a detour into the strangely loopy back-story to the translator-function, including Jacques Derrida's reflections on the ex-orbitant in *Of Grammatology*, Roland Barthes's "The Death of the Author," and especially, of course, Michel Foucault's "What Is an Author?"

J.L. Austin's (1962/1975) explosively productive theory of the performative utterance has been fruitfully applied to translation—see, for example, Robinson (2003)—but in 2003 Eve Sedgwick also published a radically transformative collection of essays in affect theory that also brilliantly stirred up the theory of performativity. As the Eve Sedgwick Pandemonium (ESP) note, the Austin Pandemonium's (JLAP's) performative revolves around what "I" do to "you" with words; the ESP asks whether performativity doesn't need to be *ratified* by what they call "witnesses." It should be obvious how the "ratification" of professional norms for translators is essential to the translation marketplace, and more specifically to the formation of a normative translator-function; but could there be strange loops of translational periperformativity as well? And could they have a constitutive effect not only on the construction of the translatorial "I," as in Chapter 2, but on the target culture?

These are the questions raised in Chapter 3, "The Strange Loops of Translation as (Peri)Performative Identities," a series of engagements with essays in a collection titled *Philosophy's Treason*. We begin by considering the cognitive parallels between strange loops and other logical impasses, namely the liar paradox and rhetorical aporias, using that latter, via the question "How do we as Derrida's readers contribute to the viability of a Derridean aporia?", to segue into section 3.1, "Logical Aporias and the Strange Loops of Periperformative Workarounds." There Mauricio Mendonça Cardozo walks us through Fleck's (1986) "exoteric circle" as applied to the circulation through translation studies (TS) in-groupers and out-groupers, "inside translation studies" and "outside translation studies"—those relentless border-crossings that many puristically minded in-groupers lament as dissipative for disciplinary coherence and others (including Mendonça Cardozo and the author of this book) celebrate as generative for knowledge-exchange.

In section 3.2, "The Strange Loops of Translating Heidegger's Untranslatables," we engage the knotty issue of untranslatability, which is not only the negative pole of the normative translation ideal of perfect reproduction—whatever is not perfectly translatable is untranslatable—but also a strategic negation of pragmatic workarounds for translatability. The argument for untranslatability tends to be purely semantic, in the sense of defining a source-textual term as possessing a stable semantic field that cannot be reproduced in the target language; a periperformative approach to (un)translatability begins not with abstract semantic fields in the null context but with what target readers can *understand*, and explores strategies for getting various understanding-based workarounds ratified by the target culture as translations.

In the essay by Sabina Folnović Jaitner we find the Martin Heidegger Pandemonium's (MHP's) term *Dasein* identified as a philosophical "untranslatable"—it is, after all, typically left untranslated in English versions of Heidegger's work—and to that we add another, *das Man*, which is typically translated (but, Heidegger scholars agree, badly) as "the 'they.'" Rather than simply debating effective vs. ineffective translation strategies for these terms, we attempt to slot them into the (peri)performative grammar outlined by Sedgwick, where the performative is what "I" do to "you" with words and the periperformative is how "they" shape and ratify the performative: the "I" there would be *Dasein* (thus "What I Do"), "you" is the world or the individual, and the "they" would be *das Man*. It should be clear, of course, that *das Man* is actually "one" or "the 'one'"—the impersonal singular third-person pronoun—and we argue that "one" is actually a more accurate pronominal figure for Sedgwick's concept of periperformativity (thus *das Man* as "What One Does"). As the argument develops, Heidegger's philosophy itself becomes an extended and often disturbing engagement with periperformativity.

In section 3.3, "The Strange Loops of 'Good' and 'Bad' (Periperformative) Translatabilities," we examine Tatevik Gukasyan's English translation of Natalia S. Avtonomova's Russian article draft as a test case for the periperformative crowd-sourcing of translation quality assessment. Given that translation is normatively defined as the exact reproduction of the source text in a target language, and that failure to achieve that impossible goal is normatively assigned the opprobrium of "non-translation"—failure to translate—the only way any translation can be judged any shade at all of "good" is through periperformative crowd-sourcing. This section explores the strange loops of such up-or-down judgments.

Finally, in section 3.4, "The Strange Loops by which Translation Shapes Collective Subjectivities," we set the DHP in dialogue with the Sakai Naoki and Lydia H. Liu Pandemonia in order to explore more fully the strangely loopy translational construction of culture—the target culture, first and foremost, but specifically the local culture *as* a target culture, which is to say as a distinct and unique and coherent culture that can stand in a complex reciprocal relation with another distinct and unique and coherent culture defined as "the source culture." This investigation mainly entails a strange-loops reading of Sakai's (1997) argument that eighteenth-century Japanese intellectuals constructed "Japan"—Japan as a single coherent national culture, Japanese as a national language—in cofigurative engagement with China/Chinese and the West.

Chapter 4, "The Strange Loops of Translational Bodies," explores the strange loops of embodiment in translation—not just the shaping and guiding of cognition by and through affect, but the impact on translation of the mouths and ears and rhythms of orality, even in written texts. After a new strange-loops pass through *The Translator's Turn* (Robinson 1991) in section 4.1, we dive in section 4.2, "The Strange Loops of Knowledge-Translation as Mouthable Rhythm," into the most radical theory of translational embodiment we know, Henri Meschonnic's insistence that we translate for the mouth and the ear, mobilizing not only "mouthability" and the *teʿamim* added to the Hebrew Bible by the Masoretes to indicate how the text should be read aloud—the singular form of *teʿamim* being *taʿam* or "taste in one's mouth, flavor"—but the intersubjectivity of serial rhythms.

Section 4.3, "The Strange Loops of the Translator's Constructivist Agency," engages the phantasmatic claims about social-constructivist approaches to translation launched by the South African scholar Kobus Marais (2014). The problem faced by the Kobus Marais Pandemonium (KMP) is akin to the problem faced by the Douglas Hofstadter Pandemonium (DHP) in claiming that after several years of consciousness-creating/-perpetuating/-perceiving strange loops the "I" doing the perceiving somehow "locks in" and comes to seem real: how does this happen? Is it a conscious cognitive decision? If so, then perhaps the KMP are right that the only people who might be tempted to imagine such control over reality would be the Western descendants of colonizers. The fact is that no one actually makes that claim—the KMP fantasize it—but the field is left open to such imaginings, by the DHP as well as the KMP, by the fact that social constructivists don't explore the socioaffective neuroscience of that "locking in." This section tracks the answers neuroscientists are coming up with to those questions.

In the Conclusion, "The Strange Loops of Translation as Transgressive Circulations," finally, rather than tying down the loose ends that have popped out of the strange loops along the way, we pluck as many more strands free as we can. Specifically, the perspective here is the avant-garde poetics of the Johannes Göransson Pandemonium (JGP), as they interact and strike sparks with the DHP's claims that because strange loops aren't "real," in the way that subatomic particles are "real," the "I" that is constructed out of them is an "illusion," a "myth," a "hallucination hallucinated by a hallucination," a "trick" or a "sleight of mind" played on us by our bodies. The JGP, too, smile at fear-mongering claims that translation isn't "real," that it's a kitschy "hoax" that endangers/contaminates the true interiority of authentic poetry—and *celebrate* that ir- or surreality, not as an abstraction or other immateriality, but as abjected bodily fluids, disease, decay, and death.

I.5 Acknowledgments

This is not our first foray into the complexities of cognitive circularity: our interest in the DHP's strange loops grows out of previous publications on the double-bind (Robinson 1992, 1995, 2006, 2017c: ch. 2); the liar paradox (Robinson 2010, 2013a: 177–206); the hermeneutic circle (Robinson 2013b: 12); the cycling around of Peircean triads from Thirdness back into Firstness (Robinson 1997/2020: ch. 3, 2016d: 182–4, 236, 239); and verbs like "to heal" and "to grow" (and Aristotle's πίστις/*pistis* "persuading/being persuaded," and Mengzi's 治 *zhì* "governing/being governed": Robinson 2014c, 2016b: 62–7). Clearly, circular cognition holds a powerful attraction for us! And we would like to thank all the colleagues who helped us think through those previous thought-circles, including Martha Cheung on the double-bind, at the 2011 Translation Research Summer Seminar after we delivered Robinson (2017c: ch. 2); Paisley Livingston on the liar paradox, walking through the halls of Lingnan University; Ritva Hartama-Heinonen and Ubaldo Stecconi, in published responses to the Peircean cyclicality of *Becoming a Translator*; Radegundis Stolze on the hermeneutic circle, at a conference in Mainland China where we were both keynoters; and the esteemed Sinologists in attendance at the April 2013 conference at the Chinese University of Hong Kong, where we delivered Robinson (2014c). Double binds (p. 116), the liar paradox

(p. 101), Peircean triadic cyclicality (p. 143), and circular verbs (p. 151) are at play in this book as well, very much in passing, as if in remembrance of those earlier instantiations of circular cognition.

In February 1997, Sonia Colina invited us to Indiana University to give a guest lecture, and in connection with that visit Douglas Hofstadter invited us to attend the graduate seminar they were giving that semester on verse translation—a seminar exploring the formalist (meter-and-rhyme) and cognitivist complexities of translation that informed *Le Ton Beau de Marot*, which was then in production at Basic Books.[6] The DHP, whom we called Doug, and who called us Doug, had the publisher send us bound page proofs of the book, and we read most of it before arriving in Bloomington. The seminar ended up being one of those mind-blowingly enjoyable experiences that happen all too infrequently in academia: a relentless romp through creative renditions of various poems in various languages into various other languages. Our gratitude for that invitation and that reception warms our heart even now, more than two decades later.

Oddly, though we had read quite a bit of the DHP before cracking the page proofs of *Le Ton Beau de Marot*, we had not read a single line of *Gödel, Escher, Bach*, and so had no inkling of the explanatory power that lurked inside strange loops. That came later.

At the Chicago MLA in January 2019, we had several fruitful conversations with the JGP, during one of which they gave us a review copy of their book *Transgressive Circulation*. That same night we took it home and read almost the first half of it—it's a short book—and found our demons in a lather about the parallels between the JGP's transgressive circulations and the DHP's strange loops. It was not too many more days after that before we had written a first draft of the piece that was eventually completely overhauled for use as the Conclusion here (after various introductory paragraphs about the DHP's theorization of strange loops had been moved to the Introduction). Fortuitously, just as we were finishing that draft, we stumbled upon a tweet about a new online journal founded and edited by the Melissa Mesku Pandemonium (MMP) called ಠಠಠ, aka "many loops" (http://xn--wgiaa.ws/editors-letter). That looked like the perfect place to send the new piece to, and we did—but didn't hear back for a good six months, so, thinking perhaps the weirdly wonderful new journal had gone belly-up, we sent it to the Brian Baer Pandemonium (BBP) at *TIS: Translation and Interpreting Studies*. After Brian had accepted it, Melissa wrote saying that they absolutely loved the piece and wanted to publish it—so we sent them the new Introduction instead, which they were also happy to have. Thanks to the JGP, the MMP, and the BBP for inspiration and contextualization.

One last professional acknowledgment: after a productive conversation on the strange loops of translation at a conference in Hong Kong with Sandra Halverson, who had read and loved *Le Ton Beau de Marot* but (like us in 1997) never heard of strange loops, we were invited by Sandra and Álvaro Marín to submit a paper to the panel they were proposing for the Genealogies of Knowledge II conference (meant to be held in Hong Kong, April 7–9, 2020, but canceled due to the COVID-19 pandemic), "Contesting Epistemologies in Cognitive Translation and Interpreting Studies: The Current State of Play." The abstract we submitted, titled "The Strange Loops of Translation," drew on drafts of this book: thanks to Sandra and Álvaro for the invitation.

The writing of this book was interrupted for the better part of six months by several big things, including the translation from Finnish of a 450-page memoir, a massive update of *Becoming a Translator* for the fourth edition, and our wife Sveta Ilinskaya's completion, submission, defense, and revision of their Ph.D. dissertation. There is no worthier project to interrupt the writing of a monograph than that last! Congratulations and thanks to Sveta, and to our daughter Agnes, who didn't see their mother much for the year or so leading up to final approval.

1

The Strange Loops of (Non)Equivalence

One of the DHP's recurring themes in "Translator, Trader" (Hofstadter 2009) is the old one, the standard one that has dominated thinking about translation for over two thousand years: translate sense for sense, not word for word. Given the thousands of writers and translators who have weighed in on this topic since Cicero first launched it in the first century BCE, one might think that would be a safe topic. Interestingly, however, the DHP demons get themselves into hot water with it.

The problem, we submit, is that the DHP identify a translational "loop" that lies at the core of human cognition (Hofstadter 2009: 37), but don't bring their usual cognitivist brilliance to bear on it. Not only is the loop as they conceive it not strange; it seems to be vulnerable to complete and utter breakdown, cognitive collapse, not through full-blown psychosis or senile dementia, but through *word-for-word translation*:

> Words trigger thoughts, and thoughts trigger words. This little loop is really the most central facet of what makes us human beings. If you eliminate the thought part of the loop, though, and insist that in translation, *words* must trigger *words* without *thoughts* intervening, then you are denying what makes us human, and any translation based on such a philosophy will turn out arid and wooden at best, and absurd and incomprehensible at worst.

What shall we do with that? We propose to read the passage, and others like it, first as a precognitivist/commonsensical reductivism, then as the strange loop that the DHP neglect to find in it.

1.1 The Campaign Against Word-for-Word Translation

What the DHP are explicitly saying in that passage is that literal translation is a matter of mindlessly, mechanically replacing word after word in the source language with their exact literal equivalents in the target language, in exactly the same order; and because that mechanical processing is *literally* and therefore nightmarishly mindless, because in being forced to translate literally the translator is *blocked from all cognitive activity*, imposing a prescriptive literalist norm on translation "den[ies] what makes us human."

But let's refine that claim a little. What the DHP's hypothetical literal translator is presumably doing is actually "mechanically" replacing each word in the source language not with its "exact literal equivalent in the target language" (such things are extraordinarily rare) but with its most "obvious," and therefore worn, tired, hackneyed equivalent in the target language. One common way of describing this kind of choice is that it's the first target-textual word in the entry for a given source-textual word in a bilingual dictionary. This would be the presumptive etiology of the normative "dumbing down" of "all" translation—the ostensible "translation universal," in Pym's (2010: 79) summary of Blum-Kulka and Levenston (1983: 119), that "translations tend to have a narrower range of lexical items than do non-translations, and they tend to have a higher proportion of high-frequency lexical items." The result of this lexical simplification, Pym continues, summarizing Toury (1995: 268–73), is that "the language [of translations] is usually flatter, less structured, less ambiguous, less specific to a given text, more habitual, and so on" (79).

"Tends to," "usually": this is not, obviously, a universal. The "I"-demons of the DHP would no doubt protest vociferously that they at least do not translate in this "mindless" way. It is, however, common enough to give the *impression* of universality. The idea would be that translators "mechanically" grab the first word that occurs to them and slap it down without considering the larger cotextual environment in which it appears. We would guess that this is the "mindlessness" against which Hofstadter rails—and quite rightly, we think, though problematically, since it isn't really mindless. It's only habitualized, so that it *feels* mindless, because it feels mechanical; and it feels mechanical because memory and nuanced analytical contextualizations are working in background mode, without conscious analysis or interpretation, let alone literary heightening.

The danger in the DHP's wild exaggeration—habitualized thought as no thought at all—is that it pushes them to jump to cognitively invalid conclusions. Because literal translation banishes *thought*, they think, in requiring literal translation "you are denying what makes us human." We imagine that as a cognitive scientist the DHP would agree that habitualized thought is quite human;

without it, in fact, the "mindless" routines we practice every day (brushing our teeth, tying our shoes, etc.) would come to feel, well, dehumanizing.

The dire warning that word-for-word translation *requires and allows* no thought is something of a bromide, of course. Translation instructors, historically tearing out their hair at their students' apparently mindless inclination to translate word for word, have launched many an equally (apparently) mindless scare tactic to impress upon them the importance of thinking more deeply and complexly about the meanings of words, phrases, clauses, sentences, and so on. What is surprising about the warning coming from someone like the DHP, however, is that it is—only ostensibly? only superficially?—embarrassingly precognitivist. The notion that it is even *possible* for words in a language a human being understands to "trigger" in that same human being other words "without *thought* intervening" is based on an absurd assumption about language and thought that no cognitive scientist—certainly not one of the DHP's caliber, certainly not a brilliant cognitive scientist who has done a lot of translating and written wittily and thoughtfully and extensively about translation—could ever possibly countenance. In Google Translate, yes, of course—an algorithmic imitation of human thought—"*words* [do indeed] trigger *words* without *thoughts* intervening." But not in human cognition. The notion that translating word for word not only *requires* no thought of the translator but *allows* none is too wildly absurd for us to imagine any cognitive scientist seriously contemplating it—especially one who has spent even a modicum of time translating. The absurdity is rather like the Noam Chomsky Pandemonium, the world's most famous pandemonial linguist, apparently actually believing that "Colorless green ideas sleep furiously" is made up of English words but is nevertheless a "meaningless" sentence. (Have they never read a modern poem?)

> *Excursus on Google Translate*. The DHP (Hofstadter 2009) invoke Google Translate as precisely an exemplar of translation with "zero intelligence" (69) "on the basis of image-free words" (86). While true enough in a sense, however, that is also a problematic claim—and therefore quite interesting.
>
> At a translation conference in Kunming, Yunnan Province, PRC, the Cameroonian translation scholar-teacher Kizito Tekwa (2019) gave a talk on the difficulties they have had teaching their Chinese students at Shanghai International Studies University to take a certain thoughtful distance from the lexical surface of the source text. Because the learning strategies instilled in them by the Chinese educational system tend to be memorization-based, they typically render the source text literally—the old complaint about translation

students, which (perhaps unwittingly) the DHP repeat and escalate to an accusation of complete inhuman mindlessness. To break them of this habit, at first the Kizito Tekwa Pandemonium (KTP) did what translation instructors invariably do when faced with this apparent "mindlessness" against which the DHP, too, rail: advised them to step back from each sentence and think about it, figure out what it meant, and translate that rather than the words on the page. When that advice, despite being delivered repeatedly over the course of a whole year, unaccountably had no effect, the KTP did something unusual: rather than simply railing at (or about) the students' refusal or inability to learn, they shifted tactics. They tried reading the source text aloud and asking the students to take notes, as if they were interpreting consecutively, but then asking them to write down their translation rather than speaking it. (The KTP call this method "interpretranslation.")

This teaching strategy took the students out of their comfort zone, which was of course a good thing for learning purposes, but also tended to increase stress; so the KTP decided to try another method alongside interpretranslation, namely post-editing machine translation. Instead of translating the written source text directly, they asked students to run it through Baidu or some other free online MT app, and then to (a) list the typical errors the MT program made, (b) correct the errors identified in a, (c) identify moments in the MT output that were syntactically and semantically correct enough but did not take cultural differences into account, (d) rewrite the passages identified in c, and (e) write a reflection on the experience.

In their paper the KTP then showed the results of a study they had done comparing the processes and products emerging out of the two approaches. What we found most interesting, however, was the *similarity* between the students translating literally based on memorization and a free online machine translator applying neural MT strategies to the source text. The KTP did not dwell on that similarity, but it took shape in our imagination as two parallel two-step processes:

The significant points to underscore there as a riposte to the DHP is that both (a1) the students translating habitually and therefore (too) literally and (b1) the MT program translating bodilessly/worldlessly and therefore (too) literally are (a2/b2) *thinking*, even if inadequately; and that the (a3) interpretranslation and (b3) MT post-editing exercises the KTP has introduced into their classes are aimed at *enhancing* students' ability to "think"—which is to say, their ability to think more consciously and complexly about the words on the page and the unstated nuances "behind" them. This (a3) "enhanced thought" would obviously be the ideal goal of training in sense-for-sense translation.

a. Output of students translating literally	b. Output of MT
1. Based on dictionary work, the students store individual words (and perhaps phrases) and their target-language equivalents in memory; that memory is habitualized, and comes to seem mechanized.	1. Based on bilingual corpora, the MT program constructs trajectories from source to target and registers those trajectories in memory for future use; that memory is in a computer, and so seems mechanized.
2. Because the students are humans, however, they do not actually mobilize their "mechanical" memory mechanically: they have to adapt the dictionary equivalencies they have memorized contextually (but the resulting output is still *too* literal for the instructor).	2. Because the MT program is a neural net, it can learn, and as a result will not mobilize its "mechanical" memory mechanically: it too will adjust its translations contextually, based on two-word, three-word, four-word (and so on) patterns it recognizes in the source- and target-language corpora (but the resulting output is still *too* literal for the instructor).
3. By stretching and enhancing the gap between a1 and a2, the students learn to raise the unconscious thought processes at work in a2 into conscious engagements with the underlying layers of meaning in whole sentences, paragraphs, and texts.	3. By not only post-editing MT output but itemizing, classifying, and correcting its errors and then reflecting on that process, students learn to raise the thought processes simulated in b2 into effective cognitive support for a3.

After the KTP (2019) came a presentation by 戴光荣, Dai Guangrong. Professor Dai (2019) followed Lynne Bowker (2000, 2001) and Juliane House (2005) in advocating for the use of bilingual corpora in teaching translation, translating, and evaluating translation quality. The juxtaposition of the two papers was striking: both the free online MT apps and the students and instructors in the translation classes taught by Dai were using bilingual corpora to *standardize* translation quality, by taking the "human element" out of the process; but also, presumably, like the KTP's students engaged in (b3) post-editing the MT output, Dai's students were also (≈b3) mentally transforming the bilingual corpus output into acceptable target-language sentences. In other words, Dai's talk also underscored the desirability in translation (and especially translator training) not only of computer-aided standardization but of smart human adjustments made to that standardization.

Now, Hofstadter (2009) was written before Google developed neural machine translation (NMT), but even the statistical machine translation architecture that Google was using back then was arguably less mindless than the DHP wanted to believe. The use of (b2) statistical frequencies for various interlingual equivalencies to guide translational choices was specifically a simulation of (a2) the human thought processes the KTP's Chinese students used to adapt the dictionary definitions they'd memorized to the specific cotextual environments in which they were expected to produce translations.

Unlike the rigid binary the DHP project between *mindless machine* and *human thought*, in other words, the word-for-word vs. sense-for-sense divide is scalar: the sense-for-sense people (the majority) insist that word-for-word translation is *too mechanical* and *not thoughtful enough*, and therefore *seemingly* mindless. As the KTP's pretty standard advice to their students showed, the push toward more sense-for-sense translation is a push for (a3) more conscious and more nuanced thought about the deeper meanings of source-textual sentences and the different ways the target language might prefer to express those meanings. *End of excursus.*

Following up on the phenomenological reading of the DHP's exaggerations—assuming that they don't really mean "without thought" but rather the *feeling* of "no thought"—can also help us make sense of a later claim in Hofstadter (2009: 64): "Were I told that I had to adopt the principle of such rigid 'faithfulness' to the author, then I would just give up translating, for it wouldn't allow me to use my own mind. It would turn me into a dull automaton, and it would remove all joy from the act." If we are unkind enough to read that literally, they are saying that following a rule of slavish fidelity in translating would *deprive them of the use of their own mind*. Following a rule rules out thought. That is patently absurd. But what if what the DHP's authorial "I"-demons really mean here—reading thoughtfully past the mind-bogglingly mindless words about mindlessness—is not that translating slavishly would literally "turn [them] into a dull automaton," but that it would *feel* to them as if following that rule would limit their expressive and imaginative freedom excessively, *as if* turning them into a dull automaton. As we'll see in Chapter 4 (p. 138), the DHP demons are ambivalent about the phenomenology of feeling—they mostly ridicule it—but it's hard to imagine them here dismissing a phenomenological reading of the feeling of self-automation, especially as they then immediately add that "it would remove all joy from the act." Joy is a feeling; and the removal of joy is also a feeling. (The removal of joy is not a removal of feeling. It's a replacement of a joyful feeling with an unjoyful feeling. There is a joy that creative people tend to feel at the ability to express themselves freely; and there is a discomfort, or even a mad ache, that those same people feel at the forcible curtailment of their expressive freedom.)

This would be the translator's expressive freedom as constitutive not only of the "I" but of the *feeling* of the "I"—the feeling that the "I" exists, is real, is alive, and is wonderful.

As any cognitive scientist is surely well aware, of course, following a rule always requires not only thought but creative thought, because real-world conditions always passively but doggedly resist rule-following. Because literal translation is almost never possible in any "pure" or "perfect" sense, *performing* literal translation—performing a not-quite-literal translation *as* a literal translation—is an act, a staging, a *mise-en-scène* that is perhaps empirically impossible, but for that very reason invariably also creative. The divergent syntactic, semantic, pragmatic, and cultural inclinations of any two languages— say, a source language and a target language—will always put up resistance to "equivalent" "reproduction" at any level, from the most slavishly literal to the most radically transformative. Translators of technical texts and back-translators tend to gravitate toward the slavish end of that cline, but that emphatically does not preclude creativity (see, e.g., Robinson 1998). It may *feel* like mindlessly uncreative work to translators of literary and advertising texts, but that is a situated affective phenomenology, not an empirical lack of creative thought.

To the absurdity of the claim that literal translation bypasses "thought," a translation scholar would add three additional problems, to which we will return later.

Problem 1: the flattening, generalizing, and habitualizing of text that the DHP hyperbolize as using words without thinking about them affects sense-for-sense translation just as devastatingly as it does word-for-word translation. A generous reading would find this observation implicit in the "at best" part of their claim that "any translation based on such a philosophy will turn out arid and wooden at best, and absurd and incomprehensible at worst"—except that it's not the *philosophy* behind sense-for-sense translation (which the DHP themselves champion) that causes the "aridity and woodenness," but the habitualization of cognitive and affective language processing.

Problem 2: the habitualization of cognitive and affective language processing doesn't always produce "arid and wooden" translations. Our textbook *Becoming a Translator* (Robinson 1997/2020) is based on the observation that, while translation is always a creative and intelligent activity, experienced technical, commercial, medical, and legal translators often habitualize transfer patterns and lexical memory to the extent that they can produce lively and accurate translations quite quickly and quite enjoyably—so long as they don't run into trouble with unfamiliar or badly used terms or phrases. When a problem like that arises, they have to stop, climb up out of habitual language processing ("the autopilot"), and deal consciously and analytically with the blockage. But this does not mean that they are not "thinking" while in "autopilot" mode. It means that their thinking is proceeding, often extremely effectively, and even creatively, in background mode.

Problem 3: deliberate literalism, highly stylized literalism as an estranging literary effect, requires intensified creativity—creativity channeled not only through "thought," but through complex socioaffective ecologies of nuance—and the result, assuming that the "philosophy" of literalism comprises the "at worst" part of Hofstadter's attack, is emphatically not "absurd and incomprehensible." As we'll see in section 1.4 below, stylized literalism has been championed by major thinkers on translation as the "best" or the "only authentic" way to translate—with, perhaps, problematic checked baggage stowed in the hold. And radical literalism has been variously mobilized for the creative purposes of highlighting the affordances of oral poetry (ethnopoetics), exploring the cross-lingual possibilities of the sounds of source-textual syllables (homophonic translation), translating nonsense poetry like заум/*zaum* "beyonsense," and so on.[1]

The best one can say here is that the DHP "I"-demons probably don't know enough about the history of literary translation to be aware of such transgressive experiments.

1.2 The Strange Loops of Sense-for-Sense Translation: Jerome

But now let us begin expanding the DHP's problematic pronouncements on translation by exploring the strange loops that they are neglecting to explore (or perhaps even to notice).

Indexicals—"this," "here," "I," and so on—are the first telltale signs of self-referential strange loopiness in the DHP (Hofstadter 2007: 160), and, perhaps not coincidentally, they are extremely problematic for translators as well. Hofstadter (2009) may have forgotten or neglected to mention this, but that earlier instantiation of the translation-theorist DHP (Hofstadter 1997: 443–9) knew it quite well: they gave there an example in which Rudy Kousbroek's Dutch "*dit pangram bevat vijf a's*" would be translated literally as "*this pangram has five a's*" (443), but in fact (a) the English pangram in question only had *four* a's, and

Lee Sallows translated it that way, and (b) "There is a certain peculiar philosophy of translation—I will call its adherents the 'rigidists'—that would insist that Lee Sallows' English pangram fails utterly as a translation of Rudy Kousbroek's Dutch pangram, but for an altogether different reason—namely, because *the two sentences talk about different things*" (445). For the rigidists, Hofstadter says, the correct translation would have to be "*Rudy Kousbroek's pangram contains five a's . . .* " (445).

This discussion veers tantalizingly close to the insight that the translation of a self-referential indexical like "dit" or "this" is a strange loop. The DHP demons sidestep that potential collision by "solving" the conundrum: what the translator *should* translate is not the literal meaning or surface referent but the implied focus or force of the word or phrase. The DHP's "rigidists" are dogmatically insisting on translating word for word, and that insistence seems to trap translators and their readers in logical paradoxes, but never fear! If one translates sense for sense, translate what is *implicitly meant*, one avoids the paradoxical loops.

Unfortunately for this "solution," things are rather more complicated than that.

As we've been seeing, the "solution" invokes the principle that translators and commentators have been telling their readers and students for a very long time—about twenty centuries, in fact, if we date it back to Cicero in 55 before the Common Era: "And I did not translate them as an interpreter, but as an orator, keeping the same ideas and the forms, or as one might say, the 'figures' of thought, but in language which conforms to our usage" (Robinson 1997/2014: 9). The *term* "sense-for-sense translation," however, was coined about sixteen centuries ago, by Jerome in their letter to Pammachius in 395 of the Common Era: "Now I not only admit but freely announce that in translating from the Greek—except of course in the case of Holy Scripture, where even the syntax contains a mystery—I render, not word for word, but sense for sense" (Robinson 1997/2014: 25).

What most mainstream translation scholars have always read there is that last part of Jerome's declaration: "I render, not word for word, but sense for sense." A skeptic inclined to challenge the DHP's (or anyone else's) serene reduction of translational strange-loops complexity through the invocation of Jerome's sense-for-sense translation, however, might want to highlight the "exception": they *always* render sense for sense, they boast, "*except of course*," they waffle, "in the case of Holy Scripture, where even the syntax contains a mystery."

Two different text types, right? Scripture and everything else—and one translation strategy for each, word-for-word for Scripture, sense-for-sense for everything else.² Since Françoise Sagan's *La Chamade/That Mad Ache* and Rudy Kousbroek's Dutch "*dit pangram bevat vijf a's*" both fall squarely into the catch-all category of "everything else," we can safely rest in Jerome's admission/announcement that they translate sense for sense. Right?

If y'all are happy with the reductivism that produces that binary, sure, why not. The awkward fact, though, is that the binary doesn't even work for Jerome's ESHP demons themselves.

The letter to Pammachius is the ESHP's defense of their sense-for-sense translation of a letter from Pope Epiphanius to Bishop John—a nonscriptural text presumably not bound by their exception. In the course of their defense, however, they cite as scriptural support for their sense-for-sense strategy example after example of the Seventy and the writers of the four gospels translating Scripture into Greek sense for sense, which makes the "syntax contains a mystery" exception look a bit dodgy. If to capture the divine mystery incarnated in the syntax of Holy Writ the Bible translator must translate word for word, then the sense-for-sense Greek translations essayed of Hebrew and Aramaic Holy Writ by the Seventy and the Evangelists are not only wrong but heretical, and must be condemned—not marshalled as Exhibits A and B in the ESHP's defense of their sense-for-sense translation of a nonscriptural letter.

More than that, the ESHP, one (or more) of the few Church Fathers who could actually read Hebrew, but arguably also the Church Father whose "I"-demons are the most radically fractalized on the claims they want to make, present those exhibits *both* as defenses of sense-for-sense translation *and* as egregious translation errors. The latter defense must be conducted surreptitiously, needless to say—hinted at tonally—because the Evangelists in particular are revealing divine mysteries to us about Jesus Christ and therefore must be strictly above reproach. For Augustine, who had no Hebrew, the Septuagint Greek translation of the Hebrew Bible too had to be above reproach: their strenuous argument in *On Christian Doctrine* (397–426 CE) was that it was inspired by the Holy Spirit, three centuries before the birth of Jesus (Robinson 1997/2014: 34). For the ESHP's demons, however, the Greek Bible as it was known in the fourth century of the Common Era was (a) riddled with translation errors which they must present as (b) exemplary cases of entirely acceptable sense-for-sense translation, even though (c) they themselves would never translate Holy Writ sense for sense because "even the syntax contains a mystery."

One possible schematization of the ESHP's segmentational strange loop:

Step 1. The situation: the ESHP have translated a letter, and have been attacked by small minds for deviating from the source text.

Step 2. The first line of defense that emerges ballistically in their mind(s) is that the great classical authorities (seem to/arguably/constitutively) mandate the sense-for-sense translation strategy they employed: the Marcus Tullius Cicero Pandemonium (MTCP) describe their method along similar lines, using the trope of paying by weight rather than counting out coins; the Quintus Horatius Flaccus Pandemonium (QHFP) inveigle against word-for-word translation; on the model of Horace's coinage, the ESHP become the first to coin the term "sense-for-sense" for the strategy that they take from the MTCP.

Step 3. But then they have been working on the Vulgate Bible translation, and studying the Septuagint Old Testament translation into Greek and the Evangelists' Greek quotations/translations from the Hebrew Bible, and have spotted all the egregious errors those writers have made; perhaps those cases could be mobilized for the defense of sense-for-sense translation as well? That's a bit of a stretch, obviously—an error is an error, and the ESHP are not accustomed to looking the other way when confronted with an error—but this is the Holy Bible they're instantiating now, and the heresy-hunters in the Roman Catholic Church in the ESHP's day care rather passionately about the sanctity of the Bible, so maybe it's best to bend a little on this niggling question of errors. And besides, reading those errors as "free" or "sense-for-sense" translations might put a stopper in the whine-holes of their more self-righteous critics.

Step 4. The ESHP draft their conclusion as "Therefore I render, not word for word, but sense for sense." But then a horrible thought arises (again ballistically): what about all those other self-righteous critics who go on and on about the Mysteries of Jesus Christ, and how the Holy Ghost Pandemonium (HGP) wrote the Bible with their own mystical hand(s)? What are they going to say about this apparent blanket prescription of a sense-for-sense strategy for *all* translation? True, "I render" is not quite the same thing as "everyone should always render"; in one sense the ESHP are simply defending their own translation of a crummy letter (from a bishop, but never mind: it's not the Bible). But what if people *read* their letter to Pammachius as a blanket prescription? What if this strange loop staged for the constitution of the ESHP's own singular translatorial "I" becomes constitutive for all translators? Best play it safe; best revise the conclusion by adding an exception: "Now I not only admit but freely announce that in translating from the Greek—except of course in the case of Holy Scripture, where even the syntax contains a mystery—I render, not word for word, but sense for sense." Never mind, again, that that is not how the ESHP has been translating the Vulgate (who's going to notice?). Never mind that this exception arguably implies the requirement that the HGP, in translating the Hebrew Bible into Greek through the bodies of the Seventy a half millennium or so before, should have been translating word for word as well—indeed that the HGP's imaginary (not empirical) word-for-word translation of the Septuagint should be the very *model* of translating "Holy Scripture, where even the syntax contains a mystery."

Step 5. Don't worry. Trust people to compartmentalize. Trust each group of readers to take from the text what they need: the Mysteries-of-Jesus-Christ people get their mystical word-for-word translation; the secret pragmatists get their sense-for-sense translation; everybody's happy. After all the letter's argumentative contortions and contradictions, we're back where we started, with our various initial assumptions, which now seem to be amply and adequately confirmed. We can all relax, and rest easy. No one is pushing us out of our comfort zones.

If we imagine not only the ESHP's translation practice in the Vulgate but their strangely loopy definition and defense of sense-for-sense translation in the letter to Pammachius as constituting for the next millennium and a half of Church history what in Chapter 2 we will see the Michel Foucault Pandemonium calling their normativized regulatory "author-function," and several translation scholars calling their "translator-function," what is that functionalized "I"? What is that rhetorically constructed authorial/translatorial "self"? On the one hand, the ESHP's "I" is construed rhetorically as the most irascible, the most curmudgeonly Church Father, the one always most eager to pick a fight not only with peers but with the Greek translators and authors of the Bible; on the other, the ESHP's "I" is construed rhetorically as the human vessel and channel of the HGP's incarnation in the Vulgate Bible. By insisting on the culpable mistranslations made by the Seventy and the Evangelists, the ESHP collectively constitute their "self" or their "I" as a maverick translator and translation theorist who dares challenge Church orthodoxy and thus opens the gates for Martin Luther to overthrow the exclusive Roman Catholic hegemony of the Vulgate Bible by translating it into ordinary lower-middle-class German (and to write their own irascible, curmudgeonly "Circular Letter on Translation" in 1530). Not coincidentally for our purposes here, too, they also open the gates for the DHP to defend their own first-person-singular translatorial freedom with a roundabout religious credo: "I'm naturally inclined to turn these knobs up high no matter what I'm writing, because clarity and vividness are, in some sense, my religion. I would be betraying *myself* if I didn't allow myself to be as clear and as vivid as possible when I translate" (2009: 64). But also, by insisting on the exception in translating Holy Writ, where "even the syntax contains a mystery," the ESHP collectively constitute their "self"/"I" as the dutiful channel of that mystery, through the work of the Third Person of the Holy Trinity, the Holy Ghost Pandemonium.

But tell y'all what, let's pay no attention to such complexities. Let's ignore the pluralized and contradictory self-constitution(s) of St. Eusebius Sophronius Hieronymus, and thus of the Quintessential Sense-for-Sense Bible Translator. Let's rest on the "obvious fact" that Jerome was a single person who used singular pronouns. Let's act as if we accepted the scrubbed version of sense-for-sense translation that has been purified out of the strangely loopy dross of the ESHP's letter to Pammachius, and simply, calmly, and commonsensically recognized the importance of mindfully translating the "deeper" meaning of whole sentences rather than "mindlessly" translating individual words with the first word given for each in a bilingual dictionary, like some kind of unthinking translation machine.

And in pursuit of that idealization, let us rigorously purge our *minds*—those minds that we need so desperately for the joy of sense-for-sense translation that they must be protected against the inevitable devastation wreaked upon them by mystical word-for-word translation—of all related complications as well. (If the word order of Scripture contains a mystery, surely the mystical translation of Scripture must be word-for-word, and the translator must be an unthinking and therefore inhuman channel of the Mysteries of Jesus Christ? Would that not be the normative ESHP anticipation of the DHP's condemnation of AI literalism?)

1.3 The Shared Strange Loops of Sense-for-Sense Translation

It is often missed, but should go without saying, that sense-for-sense translation invites the target reader to accept the conventionalized illusion of perfect access to another person's interiority. Rather than mechanically and mindlessly transferring the meanings of individual words, even if the source text is incoherently written, the translator thoughtfully and mindfully transfers *what the source author really meant*. This is such a stale truism by now that we don't really need to give it any serious thought. We all know the convention, all accept it, all go along with it dozens of times every day. We can certainly rely on that as a semantic foundation for the DHP's sense-for-sense "solution" to the strange loopiness of translational self-reference, right?

To be sure, some of the DHP (2007: 87–99, 147–206, etc.) demons devote dozens of pages to a theoretical demolishing of that very convention, in order to prove that the conventionalized illusion of perfect access to another person's interiority is a myth, an illusion, a hallucination—or, more politely, an audience-effect. And we would agree: the strange loop of translating the incoherently written source text that we sketched out in the Introduction (pp. 13–16) explored the ways in which the translator's need to *pretend* to be reading accurately what the author of the incoherent source text intended solves not the strange loops of translation but only the pragmatic problem of being able to submit some kind of translation and getting paid for it. If the source author is incapable of writing a coherent source text, any target text coherent enough to pass muster as a professional translation will be a rhetorical construction and projection of authorial interiority, not an accurate sense-for-sense reproduction of that author's true intentions. Here the DHP and the DRP are in perfect accord—except, of course, for the fact that the DHP (2009) don't think to extend the DHP's (2007) critiques of personal interiority to translation.

Interestingly enough, however, in their sixteenth chapter (2007: 227–40) the DHP also mobilize other "I"-demons to explore the odd phenomenon that, when their wife Carol died, they found Carol-demons living on in their own body and mind, and themselves looking out of Carol's eyes in a photo—a blending of selves, a shared "I."

A contradiction? Or only an apparent contradiction that actually works cognitively on two different hierarchical levels? Or just the normal functioning of a pluralized pandemonial "self"?

Pandemonially as divided on this issue as the DHP, we agree with both sides of the DHP's argument: that translatorial access to authorial interiority is a rhetorical construct and that this sharing of selves happens—or at least the sharing of a few ephemeral and ballistic demons. Indeed, we theorized the latter in terms of shared qualia in Robinson (2013a: ch. 4); what we would add to the DHP's account here is that it happens not only when we share a deep love with another person but when we translate.[3] The demons that we share with the source author can feel how the source author would write this or that in the target language—even when the actual flesh-and-blood living source author is proficient in the target language and disagrees with our choices. "No, we think *this* is how y'all'd say it," we retort. The strange loops of sense-for-sense translation *blend* translatorial demons with authorial demons, so that it seems like the authorial "I" has taken up residence inside the translatorial "I" and/or vice versa.[4]

For example, in early 2019 the DRP was hired by an editor at a major New York trade press to translate a memoir from Finnish to English. The Editor Pandemonium (EP) and the Source Author Pandemonium (SAP) were both female; the book was an attempt to use the memoir genre to recuperate several exemplary women in history by both telling their stories and retracing their steps out in the world. Our initial thought, therefore, involved distance, differentiation: as a Target Author Pandemonium (TAP) in a male body we would need to be careful not to indulge, unawares, our male privilege. As we started translating, too, our sense of difference lingered: we kept noticing the author's popularizing writing strategies, the ways in which they engaged their (perhaps primarily female?) readers with minimal distance, something that we have occasionally attempted in our own writing, but never (it seems to us) successfully.

After we had translated twenty or thirty pages, however, things began to change for us. We began to feel the text differently, and ourselves differently. We began to grow into the writing style, and thus also into the author's "I." The SAP wrote very short chapters, and we would translate one chapter at a time and email them to Finland for approval; they would mark them up, disagreeing with or questioning some of our translations (too strong? too weak? you miss the point here, etc.). We would respond, and the two of us would go back and forth until we found a rendition that felt good to both of us. It turned out the SAP demons cared as much about prosody as the TAP demons did. Every single time they would made a tentative suggestion to change a sentence just slightly to improve the rhythm, the suggested change did in fact improve the sentence, and we were happy to make it. They stressed the importance of first improving and then maintaining the syntactic and rhythmic structure of a recurring exclamation. Occasionally they would insert a request to find a way to maintain a crucial alliteration—something we love to do in translating (and too often, our curmudgeonly critics complain, in scholarly writing as well)—and we'd happily accommodate them. Thus when we came upon the sentence "Kuvaaja on onneksi paha suustaan ja todella hauska" (Kankimäki 2018: 249–50)—literally "The photographer is fortunately bad of their mouth [foul-mouthed] and truly funny"—we couldn't help rendering that alliteratively as "Fortunately the fotographer is foul-mouthed and funny as fuck," followed by a parenthetical note to the SAP: *ha ha, just joking*. But they liked it, and—telling us that readers had complained that their first book was too foul-mouthed to have been written by a woman, so the SAP had toned the second book down a bit, and they missed the cussing—voted to keep the alliterative potty-mouthery. What would the EP say? Turned out their only edit was to change the f in fotographer to ph (Robinson 2020b: 221). Nice! A few chapters later the SAP write about a portrait of Duchess Beatrice d'Este (1475–97) in the Uffizi that "Profiilikuvassa hänen suunsa on puristunut vittuuntuneen-kyllästyneeseen ilmeeseen, ja silmät tuijottavat mitään näkemättä. Minusta hänen olemuksensa huutaa hiljaa: *voi saatana*" (314)—literally, "In profile her mouth is squeezed into cuntedly-jaded expression, and eyes stare seeing nothing. To me her being shouts quietly: *oh Satan*." But that literal translation obviously doesn't work. Each sentence is rocket-fueled by one of the strongest Finnish swear words: "vittuuntunut" in the first sentence comes from "vittu"/"cunt" and means "pissed off," and at the end of the second sentence "Satan" (like other words for the devil) is a surprisingly powerful curse word in Lutheran cultures from Germany to Finland. So we rendered it like this: "Her mouth is compressed into a pretty-fucking-sick-of-this-shit look, and her eyes stare blankly, seeing nothing. I think her whole being is screaming silently *Jesus F. Christ*" (282). The SAP, knowing that there is no strong equivalent to "vittuuntunut" in English, applauded our "pretty-fucking-sick-of-this-shit" expansion; and while they didn't have a feeling for "Jesus F. Christ," they liked the rhythm.

So we gradually became a translating team of sorts—a larger "we," a "we" that occupied two bodies and swapped demons. The Source&Target Authorial Pandemonium (S&TAP). The book was simultaneously being translated into seven or eight other languages as well, but the SAP couldn't read those languages, so the TAP was the translator with whom they were bonding pandemonially. We would send them ten to fifteen pages a week, a pace that felt minimal to us, if we were to complete the translation by the September deadline, but was often difficult for them to keep up with; and yet, when we said we didn't mind at all if they let two or three chapters sit for a while unread, they replied that they were also intensely *curious* as to how we had translated them.

One interesting stylistic device they experimented with in one chapter was a monster sentence with a series of present perfect tenses, which, reading the SAP's sentence as a stylistic device designed to give the reader a gut sense of the frustration they experienced that day, we quite liked, and rendered with continuous present perfects ("having found... having hauled... having witnessed... having driven... having experienced... "). The SAP, having read our translation, wrote "[Bravo! (Tämän lauseen kääntämisestä)]" (Bravo! [For the translation of this sentence]). They were happy that we'd followed their syntax so closely.

The EP in New York, however, were not so happy. They wrote in the margin, "This section was very clunky; see my edits." Their edits were an attempt to break the long sentence up into shorter journalistic sentences, without the stylistic device.

So one obvious question there would be: whose clunkiness was it? Or, if the clunkiness was only in the EP demons' collective mind, whose Pandemonium were they accusing of clunky writing, the Source Author's or Target Author's? Since the SAP had written the long sentence, and the DRP demons knew we had worked hard to follow their syntax, we could have shifted all the "blame" onto them; but, of course, the EP was reading words that we wrote, and couldn't check them against the Finnish, so from their point of view the clunkiness might as well have been our "bad" translating.

But since we in fact liked the sentence—found it not "clunky" but interestingly gnarled and twisted in mimed frustration—we were disinclined to accept the "clunky" accusation, and indeed found ourselves indignant not just on the SAP's behalf (the editor's tactless marginal comment went to the SAP as well as us) but on *our* behalf. On behalf of the shared authorial/translatorial "I." It wasn't the TAP's sentence or the SAP's sentence; it was the S&TAP's sentence.

At the same time, however, we shared Americanness with the EP, and part of that sharing was an educated American's politically correct reaction to certain phrasings in the SAP's book—especially an Orientalizing exoticism, the well-meaning but occasionally rather clueless racism of a person from a Nordic country, and we would now and then suggest a rewriting that toned down the problematic implications of, say "eksoottiset maat"/"exotic countries." The SAP would protest: to these nineteenth-century European cultures Asian and African countries *were* exotic! Yes, we would say, but y'all want to add distance between y'all's voice and theirs, to make it clear that y'all understand things better than they did. "Mutta miten?"/"But how?" Well, with scare quotes, or careful rhetorical framings like "what seemed to them like 'exotic' lands." Or we could avoid the problem by finding a more neutral phrasing, like "far distant lands."

Occasionally a line seemed to us to undermine all the SAP's feminism. This is the one that we remember most vividly: "Ymmärrän, Karen: kun vain saa olla tuon ihmeellisen, kaikkein ihmeellisimmän miehen kanssa (miten onnekas sitä onkaan), ei millään muulla ole mitään väliä" (61). We translated that tentatively as "I understand, Karen: if you can just be with that amazing, the most amazing man (how lucky one is to have that), nothing else matters" (50), but with the parenthetical suggestion that "we"—that larger two-body "we"—add a little ironic distance. The SAP didn't understand what we meant, so we suggested "*it seems as if* nothing else matters." They refused: they liked the sentence the way it was. We weren't happy about that, but didn't push further, just left it as they liked it, thinking that the EP would probably suggest an edit—but they didn't. The EP either didn't notice it or thought it was just fine. So there we were, the odd man (yes, *man!*) out.

The DHP's strange loops theory would insist, too, that in knowing interactively, collectively, what the SAP really meant and didn't spell out, in reading not so much (as) the authorial but (as) the shared authorial/translatorial mind, we are in an important sense *creating* that mind—the constitutive rhetorical effect of translation on the "I." That would effectively mean that the constitutive effect sense-for-sense translation has on the source author's mind and the translator's mind is specifically a *reciprocal* audience-effect constitutive in each direction of *someone else's* "I," and thereby implicitly comes to constitute a mutual self-constitution that is complexly recursive:

(the translator-as-source-reader)
constituting the source author's "I"
in the source language
—*in so doing also in some sense*—
(the translator-as-source-reader-becoming-target-author)
performatively becoming or occupying the source author's "I"
in some mental interlingual realm that keeps veering precipitously toward
the target language
—*in so doing also normatively (but still performatively)*—
(the translator-as-target-author)
vacating their own target-authorial "I" in order (still performatively) to naturalize
in the target reader's mind the source-authorial "I" constituted by them in the target
language *as* the source author's "I"
—*in so doing also normatively (and supremely performatively)*—
(the translator-as-nobody)
constituting and naturalizing in the target reader's mind the target-authorial "I"
as vacated, empty, nonexistent, *and so on.*

If Hofstadter (2007) would seem to insist on that mutual self-constitution of authorial "I" and translatorial "I," however, Hofstadter (2009: 65) seems to reject that implication out of hand: "It's not as if I were rewriting the plot of Sagan's novel, or changing her characters' characters," they proclaim; "I am just being myself." The "I," that strange loop, is "*just* being myself." *I Am a Strange Loop* might be read as a 400-page unpacking of just that "just"—if it weren't for the fact that the DHP nowhere explore, or even acknowledge the existence of, the strangely loopy tensions between their TAP's "just being themselves" and the TAP's shared participation in the Saganian SAP's "just being themselves"... though that would arguably be the shared participation that they theorized in Chapter 16 of *I Am a Strange Loop*. If after all the Françoise Sagan Pandemonium's (FSaP's) writerly acts in *La Chamade* help (re)constitute the SAP's "I"—the "self" that they are "just being"—one of the DHP's tasks in *That Mad Ache*, as sense-for-sense translation has traditionally been understood, is to recuperate that authorial (self-)(re)constitution by becoming it, by sharing it. We are a strange loop.

Of course, that shared authorial/translatorial "we"-like "I" is never a perfect blending of rhetorical self-constructs—any more than it is between two people who are deeply in love. A lot of work has been done by translation scholars on the translator's narratoriality (see Hermans 1996; Schiavi 1996; Baker 2000), the extent to which the translator's re-narration of the source text expresses the translator's "just being themselves." But that some sort of blending of self-constructs does happen is undeniably part of the strange-loops phenomenology of sense-for-sense translation—just as, for Hofstadter, it was undeniably part of the strange-loops phenomenology of their marriage. When the DHP "I"-demons protest that "It's not as if I were rewriting the plot of Sagan's novel, or changing her characters' characters; I am just being myself" (65), however, they are trying to split the translatorial "I" off from the shared authorial/translatorial "I," and to assert the prerogatives of the translatorial "I" as if no sharing had ever taken place—as if the translatorial "I" had become a stand-alone authorial "I" in its own right. This, we argue, would be the translational equivalent of the DHP saying of the FSaP what their friends kept saying to them *just months* after Carol died: "You can't feel sorry for her! She's dead! There's no one to feel sorry for any more!" (2007: 228). The DHP's comment on that "consolation"—"How utterly, totally wrong this felt to me" (228)—would be our translatorial comment on "It's not as if I were rewriting the plot of Sagan's novel, or changing her characters'

characters; I am just being myself." (Sagan died on September 24, 2004, *just months* before the DHP started translating them. Did they not come to feel their authorial personality stirring inside them as they translated them? Do "they" and "them" there not loop the Sagan Pandemonium into the Hofstadter Pandemonium and the Hofstadter Pandemonium into the Sagan Pandemonium in strange ways?)

> *Excursus on the "being myself" of the "I."* It is striking, certainly, that Hofstadter (2007) debunks the "being" of the "I" so fiercely and then just two years later Hofstadter (2009) seems to be working so hard not only to perpetuate the myth of the "I's" singularity and autonomy but to aggrandize it, heroize it as something that *must not be betrayed*, something that the DHP's "I" must "be," as "myself." In the course of their long assault on the "reality" and singularity of the "I" the DHP's "I"-demons cite their old friend the Daniel Dennett Pandemonium (DDP), who in *Consciousness Explained* (1991: 24) compare consciousness to paper money: "it *feels* as if it is worth a great deal, but ultimately, it is just a social convention, a kind of illusion that we all tacitly agree on without ever having been asked, and which, despite being illusory, supports our entire economy" (Hofstadter 2007: 315). And yet just two years later, "It's not as if I were rewriting the plot of Sagan's novel, or changing her characters' characters; I am just being myself" (Hofstadter 2009: 65).
>
> It's easy to invoke protocols of classical logic, which are based on the transcendental presumption that each human body has a single unified "I" that must present to the world a single propositionally unified front, and proclaim "contradiction"—but of course down here on pandemonial earth things are always considerably more complicated than is dreamed of in the idealized mansions of formal logic.
>
> The DHP, for example, in undercutting the common human assumption that "my" "I" is a stable reality, also ask in many different ways whether the "I" is singular—and end up declaring "I think I come down on the 'two me's' side" (2007: 314). A fervent advocate of the "one me" side might ask snarkily which side the other "I"/"me" comes down on—one me as well?—but the DRP protests instead: only two? Surely there are more? A few pages earlier they cell-divide into those "two me's" in order to explore the big issues in an extended dialogue—something the DHP do a lot, quite successfully—and one of the "me's" asserts that "out of intimacy, out of empathy, out of friendship, and out of relatedness (as well as for other reasons), your brain's 'I' continually makes darting little forays into *other* brains, seeing things to some

extent from their point of view, and thus convincing itself that it could easily be housed in them" (Hofstadter 2007: 291). Some of the DRP "I"-demons wonder whether those "darting little forays into *other* brains" are really made by *whole* "I's," so that, if there are only two, one stays behind to guard the fort while the other goes and lives in other brains for a while. We agree that those forays are made, but have our doubts about the twoness of the "I's" or "me's" that make them.

Applied to the DHP's strange-loops model, the DDP's Pandemonium model would mean not only (a) that the *saying* of "I" that constitutes "the" "I" is multiply sourced, which entails that the "I" is itself multiply sourced, and is presented to the world as an artificially unified singularity (*the* "I"), nor simply (b) that *some* of the "I"-demons (not the whole "I") make "darting little forays into *other* brains," but also (c) that some of the apparently locally sourced "I"-demons that say "I" originated in those *other* brains, are like other-sourced spies or moles in the ballistic constitution of the local "I."

It also means that any "contradictions" we seize upon sadly or triumphantly in the DHP's (or anyone's, including the DRP's) thought can be laid at the feet of the pandemonial structure of the "I." On the one hand, the rhetorical situation in which the "I"-demons go ballistic changes the constitution of the "I": writing to their fellow cognitive scientists, the DHP (2007) demons undermine the assumed "stable reality" of the "I"; writing to lay readers of their Sagan translation, however, the DHP (2009) demons express and apparently endorse commonsensical (which is to say ideologically normative) notions like "I am just being myself." On the other hand, "I"-demons are continually being exchanged and propelled socioaffectively, so that expressing ideologically normative commonplaces like "I am just being myself" can *feel* like the "I ... just being myself." Those collectivized/normativized "I"-demons are so well disguised as "many me's" that it can come to feel normal and commonsensical to assert the grandiose autonomy of one's "I" even while translating, even though the ideological norm for translating is the radical *subtraction* of self. *End of excursus.*

As we'll see the Johannes Göransson Pandemonium (JGP) noting in the Conclusion, in fact, this very loopy volatility of the translational hijack is precisely why translators are normatively expected—and even trained—to "disappear" from their own texts, to become mere invisible channels or transparent windows through which the TRP not only authentically experiences the interiority of the SAP's "I" but vicariously experiences their own receptive interiority as the "I" of a Source Reader Pandemonium (SRP). That is the normative vanishing act that would remove the "joy" from the DHP's translatorial "I." Fear of the joyless affect that would depress them into quitting translating if they were forced to submerge their demons' TAP "I" entirely in Sagan's SAP "I" is what makes them trumpet the "right" to "just be my[singular]self" in translating. And while the DRP have found considerable joy even in slavishly translating technical texts, we agree that translating literary texts creatively is considerably more joyful. We won't be disagreeing with the DHP here.

We might complicate their Whitmanesque self-assertions a little along the way, however. That joy of translating: isn't it partly, perhaps even mostly, the joy of participating actively, creatively, exuberantly in a *shared* activity? Even if the shared authorial/translatorial "we"-construct feels guided jointly, doesn't it feel *housed* on each side, as if it dwelt in the author's "we" for the author and (more to the point here) in the translator's "we" for the translator? When we imagine that we share an authorial/translatorial agency with a Finnish memoirist SAP, isn't that a phenomenology that we (our single/multiple mind) have iteratively generated out of our engagement with their written words? (We did meet face to face over tea halfway through the project—and then went back to interacting in writing.)

The DHP demons by contrast seem to be imposing a constrictive binary on their options: either they are singularly "just being *myself*" wholly apart from Sagan or they are surrendering wholly, slavishly, joylessly, and indeed mindlessly and post-humanly to the despotic intention of the source author or the inflexible contours of the source text. Given a Hobson's choice like that, sure, bold self-assertion seems like the self-respecting translator's only option. But isn't that binary just bad theory? (Surely the Carol-loving demons of the DHP would not apply that stricture against joining a shared authorial/translatorial "I" to the shared consciousness they experienced with Carol? Surely that sharing was mostly filled with joy?)

And to answer the question with which we began this section—is the DHP's insistence in the same book on both the egregious falsity and the joyful reality of perfect access to another person's interiority a contradiction, or only an apparent contradiction that actually works cognitively at two different hierarchical levels?—I would say that the DHP's more skeptical demons are quite right that we never have *perfect* access to another person's interiority, and therefore that their conviction that they had that access with Carol was a conviction, not a fact. To put that differently, it was a shared phenomenology generated iteratively by the strange loops of intimate social interaction. Working as hard as those skeptical DHP (2007) demons do to vacate the "truth" or "reality" of that phenomenology (to dismiss it as a "myth" or an "illusion" or a "hallucination") only seems necessary to a dogmatic objectivist for whom feelings aren't real, experiences aren't real, beliefs aren't real—and certainly the strange-loops audience-effects that generate the "I" (and the "we") aren't real. More on this as we go along.

1.4 Strange Loops of Word-for-Word Translation: Friedrich Schleiermacher

And now let's explore the other vector of the convergent/imbricative ESHP dualism, the word-for-word translation strategy that for the ESHP is divine inspiration and for the DHP is mindless mechanism, and ask:

What kinds of strange loops does literal translation provoke?

How does the strangely loopy creativity of literal translation shape the translator's intercultural "I"?

How does the strangeness of a literal translation estrange the TRP's sense not only of the target language but also of their own target-cultural "I," and their "quoting" occupation of the SAP's "I"?

That verb "estrange" begins to hint at the direction in which we want to take this: away from what the DHP calls the "rigidism" of literal/word-for-word translation in the direction of what the Friedrich Schleiermacher Pandemonium (FSP) in their 1813 Academy address on the different methods of translating call "das Gefühl des fremden"/"the Feeling of the Foreign." The FSP do hint that their preferred translation strategy, what Lawrence Venuti has been calling "foreignizing" since the mid-1980s, is a form of stylized literalism (1813/2002: 81, Robinson 1997/2014: 232 in English); but for the FSP the phenomenology ("feeling") of the approach is more important than any kind of segmentational formalism (one word at a time).

Specifically, the FSP's calmer "I"-demons begin with three different levels of felt foreign language proficiency: the frustrated beginner, who feels that nothing makes sense, everything seems like gibberish; the intermediate learner, who feels able to speak the foreign language hesitatingly, with an accent, and can read texts written in the foreign language, but without a feeling of ease; and the polyglot, who hardly even notices the phenomenological difference in ease of access between the native and the foreign language.

The beginner, they say, is useless to the translator. The debates over sense-for-sense and word-for-word translation relate instead to the other two. (We do, however, explore the simulatory value of the beginner's frustration for David Melnick's homophonic translation of part of the *Iliad* in Robinson 2022a: "Hermeneutics" sec. 2.)

The "philosophy" (to stick with the DHP's usage) of sense-for-sense translation tends to privilege target texts that TRPs read with the feeling that they were originally written in the target language; and that, the calm FSP demons say, would correspond to the phenomenology of the polyglot. The traditional sense-for-sense thinking would follow the path we explored in section 1.3: the partial fusion of authorial and translatorial demons, the polyglot TAP simulating for the monoglot TRP their own polyglot feeling of reading the source text fluently. This is more or less what Eugene A. Nida (1964, 1969) called dynamic equivalence, though the Nida Pandemonium theorized neither the partial phenomenological fusion of authorial and translatorial "I"-demons nor the simulation of fluent polyglot translatoriality for the target reader, and so made the idea of simulating the source reader's response in the target reader seem like pure mystical/wishful thinking.

The FSP's calmer demons' preference is for the awkward and hesitant phenomenology of the intermediate learner, who stumbles through conversations with a heavy accent and through readings of foreign texts with the Feeling of the Foreign. Their idea is that the translator should "nachahmen"/"simulate" that phenomenology in the target language, so that the target reader with no knowledge of the source language will still get *in the target language* that Feeling of the Foreign experienced by the intermediate learner of the source language.

While clearly the FSP demons would set themselves in vehement opposition to the DHP demons' association of literalism with mindlessness, however, it is striking that the entire tissue of the FSP's argument is in analogy—in those patterned level-shifting audience-effects that for DHP (2007) are the basic building blocks of strange loops. Ground Zero for the FSP is the analogy between the Target Reader Pandemonium (TRP) reading the translation and the Target Author Pandemonium (TAP) reading the source text; in order to give the TRP the impression that they are the TAP, the latter simulates their own felt reading experience for the former. In this way the TRP's identity is iteratively shaped through the reading of translations: they gradually become not only well read but an active and worldly contributor to the target culture. In the DHP's terms, the rhetorical construction of identity emerges first through an analogous structure on each level, between the objectifying experience of the text being read and the subjectifying meta-experience of reading oneself reading the text, and then expands further across levels, the target reader imagining themselves as the translator and the translator imagining themselves as the target reader.

"It is out of a dense fabric of a myriad of invisible, throwaway analogies no grander than these," as the DHP (2007: 149) quip, "that the vast majority of our rich mental life is built." What a strange-loops reading of the FSP's Academy address adds to DHP strange-loopism is the notion that the unconscious/unaware strange-loops constitution of the "I" that the DHP theorizes might be a bad thing, a state of being that might need to be combatted with counter-cognitivisms. The pandemonial TRP(s) of a foreign novel, say, might be so wonderfully immersed in the story that they unconsciously objectify not the book or the author or the writing experience but the characters, and further (by extension) unconsciously objectify not the translator or the translatorial interpretation or their own reading experience but the reading "I." This would be the FSP's dystopian moment. This is the state of cultural being that they seek to dislodge. Because the reading "I" is being constituted unconsciously, it is being constituted *as* unconscious, as unaware, and therefore as the passive recipient—say, hapless victim—of whatever is coming across the conduit from the foreign, from the source author, through the translator.

The crucial tipping point for the FSP would be the moment at which the TRP begins to become aware of the analogies—the point at which that "wonderful" (but passivizing) immersion in a foreign work tips over into awareness not only of the reading "I" but of that "I's" similarity to and difference from the work being read. The DHP write of the use of analogy in artistic narratives, "where, because of a strong analogy that readers easily perceive between Situations A and B, lines uttered by characters in Situation A can easily be heard as applying equally well to Situation B" (2007: 152)—*can easily be*, but, the FSP would interject, *may not be* heard that way. Sometimes we would prefer not to hear or become aware of the analogies. The DHP go on to remind us of the times when analogical parallels between what's happening on screen and what's happening in our lives disturb us—as, say, when we are cheating on our partner or our partner is cheating on us, and we are both more or less successfully suppressing open knowledge or discussion of that fact, until a scene on TV reenacts what we're trying to suppress and we both have to work extra hard to go on not-noticing and not-discussing the problem.[5] Awareness and verbal acknowledgment of the analogy would be a disruptive force in our lives. It would force us to deal openly with the problem, and we would rather not do that. That is the tipping point that the FSP seek to isolate and intensify in the hermeneutical interactions between the translator and the target reader. That tipping point is generally termed "denaturalization"— the undoing of the "naturalization" of translation as the perfect and therefore unnoticed transfer of the foreign book into the target language and culture— meaning that target readers become aware that they are reading a translation, and, ideally, that the translation is the product of interpretive work done by a specific mediatory language professional. The FSP don't call it a tipping point, and don't name that tipping point denaturalization, and don't even articulate the reasoning behind their largely unconscious gravitation toward the transformative power of that moment; our ability to name and theorize those things in the FSP's rather inchoate account comes from the history of modernist estrangement theory in Shklovsky and Brecht (Robinson 2008; 2013b: ch. 4; 2017e: 144–8), and, of course, from this application of the DHP's brilliant strange-loops theory.

Now as long as the FSP's translational foreignism is just about enhancing the target reader's awareness of translation as work, as interpretation, as transformation, modern translation scholars have every reason to embrace it. As long as it is just about enhancing the target reader's *and* translator's awareness of the analogies between what they are reading and what they are doing, and, following the DHP, of the impact those analogies have on the constitution of the "I," and, further, of the transformative impact enhanced awareness of those analogies might have on the constitution of the "I," cognitive scientists have every reason to embrace it. The problem is that for the FSP it is a lot more than those things.

It is, for one thing, steeped in the hysteria of the more panicky FSP demons, who seem to have convinced themselves that the best way to wean German readers off traditional passivizing unawareness of the foreign is to hurl exaggerated and rather silly abuse at the analogies on which it is based. They claim, erroneously, that no polyglot ever wrote great literature in a foreign language—a claim that they must know is not true—but if one did, they say, they would immediately condemn that whole effort as "eine frevelhafte und magische Kunst erklären, wie das Doppeltgehen, womit der Mensch nicht nur der Gesetze der Natur zu spotten, sondern auch Andere zu verwirren gedächte" (88)/"a wicked and magical art akin to going doubled, an attempt at once to flout the laws of nature and to perplex others" (236). And because no one can *write* well in a foreign language, it is, they insist, equally wicked to *translate* as if the source author were actually writing in the target language.

Indeed, as we showed at length in Robinson (2013b: ch. 2), all of the analogies marshalled by the panicky FSP demons are cluelessly but aggressively illogical. Not only is their pragmatic application to real-world situations empirically problematic (many writers have written brilliantly in a foreign language, for example); structurally, too, their analogies are fatally wounded. For a respected philosopher who has been described as a "master logician" who "would not have made such an elementary error" (Thandeka 2005: 199), the logical morass in which the FSP's bad analogies mire the Academy address is flatout embarrassing. Arguably even worse, however, is the fact that the FSP demons defend those wounded analogies as fiercely as a wild animal its young, heaping undeserved and wildly disproportionate abuse on anyone or anything that might be construed as detracting from the viability of their claims.

Why the hysteria? Why the panic? A generous reading of the Academy address might suggest that the FSP hadn't really worked through their thesis thoughtfully enough. They needed more time, or they needed modernist estrangement theory, or they needed the DHP's strange-loops theory, and there was of course no way to get those things under their belt before writing the address on translation, so they walked into the Royal Academy session unprepared; that nagging sense of being vulnerable before the sharp minds of the academicians—especially perhaps their old sparring partner from the planning sessions for the University of Berlin a few years earlier, Alexander von Humboldt (1769–1859)—might have frazzled the FSP demons' spirit and left them on the verge of hysteria. August von Kotzebue (1761–1819), a reactionary popular dramatist whom the FSP mentioned as present at the address in a letter to their wife (Reimer 1860: 300), was not necessarily a sharp mind, but Kotzebue loved sclerotic autocracy and was infamous for taunting both Napoleon and the Prussian nationalists mercilessly; it might have put the panicky FSP demons on edge to find this savage tongue sitting across the table from them as well.

That mention of Prussian nationalists, in fact, suggests an even more pointed explanation of the FSP's hysteria in the address: by 1813 most of their "I"-demons were zealous patriots, whose zealotry often bordered on hysteria. Not only had the 1806 occupation of Prussia by Napoleon radicalized them; the Prussian king's weakness in response to that occupation, especially the king's unwillingness to modernize the state to make Prussians better able to end the current occupation and repel future ones, had incensed them. The straw that broke the pandemonial camel's back was that in reaction to the armistice called in early June, scant days before the Academy address, King Friedrich

Wilhelm III had disbanded the *Landsturm*, the reserve militia mobilized at home for civil defense; walking into the Royal Academy to read their paper on June 24, 1813, the FSP were still seething. This disbanding was for them a grotesque assault on everything they believed, a contemptible continuation of the reactionary backlash against reform. And yet they said nothing about it in public until three weeks after the Academy address, on July 14—a long time to seethe in fury without action. But apparently there was good reason to wait, and even better reason not to have said anything at all: on July 17, the king issued an order dismissing Schleiermacher from the editorship of *The Correspondent* (where the FSP had published the attack) and requiring them to leave Berlin. This order was softened, and the Schleiermacher Pandemonium was permitted to stay; but Count Friedrich von Schuckmann (1755–1834), their boss in the Ministry of the Interior, warned them that one more attack on the king and they would be out—out of Trinity Church, out of the Ministry, out of the University of Berlin, out of Berlin entirely. This warning effectively ended the FSP's political activism.

But what does any of this politicking have to do with translation? Everything, as the FSP saw it. The Academy address on the different methods of translating is a deeply and passionately political screed. The crushing problem that demanded a concerted effort toward a solution was the French occupation—and the French were foreigners. The more cosmopolitan FSP demons loved foreign cultures, foreign literatures, loved to learn foreign languages and translate from them; but their more nationalistic demons, which by then were increasingly in the ascendancy, also realized that what left Prussians and more generally German-speaking cultures across Europe vulnerable to foreign invasion and occupation was *excessive* openness to the foreign. The political task for the FSP in the Academy address, therefore, was to placate the liberal cosmopolitans and the rabid nationalists not just in the audience but among their own demons—and they did it by carving out a tricky middle ground between complete cosmopolitan openness to the foreign and a complete nationalist embargo on the foreign.

For the FSP it is precisely because the TRP's identity is so powerfully shaped by analogical level-shifts that the "right" translation strategy is so important. A "dumb" or passivized TRP who flees pressures toward translational self-awareness back into unconsciously calm self-protective objectification—"I" am a human ready-made reading this book—is left at the mercy of the foreign, which sneaks its insidious "I"-demons into the German reader's Pandemonium under the cover of the foreign author saying "I" in fluent German. It's only if the target reader is progressively made more aware not only of the foreignness of the book they're reading, but of the translatorial *madness* of that book, that the reader can fashion and mobilize protective defenses against the foreign. It's only, in other words, if the target reader can experience the foreign book through the felt analogical filter of the translator's foreignizing simulation that they become able to appreciate the foreign text without being damaged by it. The desideratum is the shaping not only of individual minds and hearts and identities but of the whole culture of German-speaking peoples in and by and through translation.

That filter for the FSP was stylized literal translation, which they believed offered the best available protection against the importation of foreign works through the Feeling of the Foreign—that affective *signal* of the analogies between reading the source text and reading the target text that, by enhancing the German target reader's uncomfortable awareness of the transformative impact those analogies would have on the individual and collective German self, could erect an effective barrier against the virulent body and spirit of foreignness. Through literal translation Germans could continue importing foreign works into their culture, and so continue to enjoy their superiority over the protectionist French, while also protecting themselves against the alienating contaminations of rootless foreign (especially French and Jewish) modernisms.

Here, however, is where strange loops begin to figure in the FSP's imagination. The fact that by morphological extension "das Gefühl des fremden"/"the Feeling of the Foreign" is also potentially a feeling of "Entfremdung"/"alienation" or "Verfremdung"/"estrangement" might also suggest that some of the FSP demons might be sneakily looking to *enhance* feelings of alienation or estrangement in German readers of translations—to make them feel *more* uprooted, more outcast, more abject—and nothing could be further from the truth. The strange-loops phenomenology of "das Gefühl *des* fremde*n*"/"the Feeling *of* the Foreign" was designed to alienate German readers of translations *from* the foreign—or rather, from the direct, unconscious, and therefore unimpeded experience of the foreign, and thus from its power to enhance alienation. In simulating the alienating force of the foreign, the Schleiermacherian translator was supposed to *abrade* the foreign speaking voice, and so to *mark* it as alien, as a kind of stylized homeopathic dose of alienation that would protect the rooted German Volk against uprooting.

The desired effect would be a retrofitting of rooted belonging as an alienation-like affect, rootedness staged phenomenologically as simulated rootlessness, yielding an aestheticized simulacrum of cosmopolitanism that would ward off uprooting. Effectively, as we say, this entailed reading the foreign text with a heavy German accent, so that one would *Feel the Foreign in the German body*, which seemed somehow "naturally" built to maintain suspicious defenses against foreign ways—seemed, in other words, to have been supplied to the German Volk by God as a kind of mystical armor against the French, the modern, the urban, the capitalist, and the Jew.

For the DHP, y'all will recall, the objective correlative of word-for-word translation was Google Translate. But if we're right in section 1.3 in reading their account of their loving connection with Carol as a marital analogue of sense-for-sense translation, where the author and the translator partially fuse into a single shared consciousness-construct that *knows* the true target-language equivalent of each source-textual word and phrase, the marital analogue of word-for-word translation in the FSP's account of the Feeling of the Foreign should really be something like the old married couple that is thoroughly estranged but not only stays together but maintains outwardly polite relations. "Hate" may be too strong a word for what they feel for each other; it's more a sardonic suspicion each feels that the *other* hates *them*, that the unfailingly polite front the other presents to them is actually a mask for deep enmity. They stay together not out of old habit, but rather so that every time they feel bitterly humiliated by the other's imagined hatred and surreptitious future victimization, the pain will remind them not to let down their guard, not to give in to pity (let alone compassion). The other is the enemy, and because it's essential not to let even the tiniest shred of love sneak in when their backs are turned, each holds the enemy close.

This is quite a different staging of literalism than the DHP's, obviously. For the Hofstadter Pandemonium, the results of literal translation are "absurdity and incoherence"; for the Schleiermacher Pandemonium, the results are equally negative, but intentionally so:

> Wer wird sich gern auflegen, in minder leichten und anmuthigen Bewegungen sich zu zeigen, als er wohl könnte, und bisweilen wenigstens schroff und steif zu erscheinen, um dem Leser so anstößig zu werden als nöthig ist, damit er das Bewußtseyn der Sache nicht verliere? Wer wird sich gern gefallen lassen, daß er für unbeholfen gehalten werde, indem er sich befleißiget, der fremden Sprache so nahe zu bleiben, als die eigene es nur erlaubt[?]

(81)

Who would publicly cripple his own verbal facility and grace in order to appear, at least at times, churlish and clumsy, and as offensive as is necessary to keep the reader aware of what is going on? Who would gladly be thought a bungler just because he took pains to stick as close to the foreign language as his own would allow[?]

(232)

Deliberate verbal self-crippling, a deliberate offending of the TRP, "damit er das *Bewußtseyn* der Sache nicht verliere," that they might not lose *awareness* of the thing. For the FSP literal translation—"der fremden Sprache so nahe zu bleiben, als die eigene es nur erlaubt," sticking as close to the foreign language as one's own will allow—is a conscious translatorial speech act with a very specific but complexly layered perlocutionary effect: (a) Feeling the Foreign so as to (b) keep it at bay, thereby (c) protecting the purity of the *Heimat*/homeland. That—(a), or (ab), or (abc)—is the "thing" of which the TRP must remain aware. So long as (b) and (c) follow "naturally" (normatively) from (a), awareness of (a) is enough. It's only when (a) triggers the polyglot's cosmopolitanism and the foreign languages and cultures begin to wield an attraction similar to, or even stronger than, the homeland's that foreignizing translation needs to escalate its illocutionary force, and begin to trigger enhanced awareness of (b) and (c) as well.

Where are the strange loops in this phenomenology of translational enmity, then?

To the extent that we want to say that any kind of enmity is a vicious cycle, of course, there is a strange-loops phenomenology to it. If by hating "you" "I" not only make "you" hate "me," but make "you" construct "my" "I" as an enemy, and if the only reason "I" hate "you" is that "your" hatred for "me" has made "me" construct "your" "I" as an enemy, and so on, around and around, each of us is simultaneously and serially both the perpetrator and the victim of enmity. And if by hatefully holding "you" close "I" not only construct "your" "I" as a living, breathing reminder of the necessity of constant vigilance against lowering "my" guard, but make "you" construct "my" "I" as a reminder of the same need for vigilance, and if the only reason that "I" am constantly invading "your" space and disrupting "your" peace of mind with the blankness of "my" face and words is that "you" won't leave "me" alone, "you" keep inciting "my" vengeful wrath against "you," and so on, around and around, each of us is simultaneously and serially both a reminder of the need for vigilance and disturbingly vulnerable to a collapse into empathy and compassion.

More than that, this kind of mutual hatred and mistrust would tend to construct not just each individual "I" as the enemy and the victim of the other, but the collectivized "we" of the two as enemy-and-victim, a kind of spinning Daoist fish-diagram of bitter victimage (victimizing/being victimized), so that every aggressive act that self-victimizes comes to be experienced as a desperate and frustrated act of self-protection and every defensive act that self-protects comes to be experienced as an angry and indignant act of self-victimization.

But how does that work in the translational simulation of the Feeling of the Foreign? "The Foreign" isn't exactly an agent with the power to *decide* to incite fear or resentment. That agency has to be projected onto "the Foreign." But then foreignizing translation is one of a very large collection of xenophobic speech acts that, by making "the Foreign" ugly, also project danger and threat onto that distant imaginary, and so bring it imaginatively close and give it imaginary agency. The very act of intensifying the feeling of needing to guard against attack escalates the fear of attack, which intensifies the need to guard against it, and so on.

Whether that strange-loops phenomenology of enmity-and-victimization is an accurate or useful representation of the Romantic Nationalist FSP demons' protectionism is one question; another is whether the liberal cosmopolitan FSP demons' conversion of that phenomenology into the more positive image of *Weltbürgerschaft* is also strangely loopy. Here, for example, is a passage from the Academy address's peroration:

> our respect for the foreign [*(unsere) Achtung für das Fremde*] and our mediatory nature [*(unsere) vermittelnde Natur*] together destine the German people to incorporate linguistically, and to preserve in the geographical center and heart of Europe [*im Mittelpunkt und Herzen von Europa*], all the treasures of both foreign and our own art and scholarship in a prodigious historical totality, so that with the help of our language everyone can enjoy, as purely and as perfectly as a foreigner can, all the beauty that the ages have wrought.
> (Schleiermacher 1813/2002: 92 in German, Robinson 1997/2014: 238 in English)

What is going on there?

At the very least, the image of "cosmopolitanism" as the amassing of "all the treasures of both foreign and our own art and scholarship in a prodigious historical totality" in the German language, with the foreign art and scholarship filtered through literal translation, seems a bit questionable. Were that pretty "cosmopolitan" state of affairs to come to pass, foreigners would not need to go to the source cultures of the works Germans had translated, but could simply move to "the geographical center and heart of Europe," learn German as best they could, and then "enjoy, as purely and as perfectly as a foreigner can, all the beauty that the ages have wrought." Presumably, since those foreigners would not be able to read German all that "purely" or "perfectly," they would also be reading the works collected in German through the filter of the Feeling of the Foreign, and for Schleiermacher that would almost certainly be a plus.

But notice also the phrase that we've translated as "our respect for the foreign." That could mean a reverence (*eine Ehrerbietung*) for the foreign, an appreciation (*eine Würdigung oder eine Anerkennung*) of the foreign. But the FSP's noun is different: they write of "[unsere] *Achtung für das Fremde*." *Die Achtung* comes from the verb *achten*, to notice, to observe, to pay attention to; and while it does get used to mean respect or esteem, it is also used as a warning: *Achtung!* (pay attention!). The German "respect for the foreign" of which the FSP write could be a watchful respect for a dangerous animal that might pose a significant threat. The FSP demons specifically want translators to mark their translations *as* translations, with stylized translationese, so as to draw *attention* to the fact that they originate outside the German-speaking "geographical center and heart of Europe." The danger of what they call "bringing the author to the reader," or what Venuti calls domesticating translation, is that it smuggles the foreign into the heart of the target culture without warning signs. This is what all the atrocious (ana)logic in the Academy address comes down to: rather than simply coming out and saying "we have to post literalist warning signs on translations so that we can Feel their Foreignness, and pay attention to the possible threats they pose to German culture," the nationalist FSP demons take on the protective coloring of their cosmopolitan pals and create crypto-nationalist (ostensibly cosmopolitan) analogies for domesticating and foreignizing translations that don't work logically. The bad logic is a cosmopolitan smokescreen for nationalist protectionism. *Die Achtung für das Fremde* is an excellent binary gate for that smokescreen, in that it swings silently: a warning against a potential danger that sounds like reverence.

The phrase that we've translated as "our mediatory nature" is also telling. Here the FSP's German phrase is [*unsere*] *vermittelnde Natur*, with *Mittel* "middle" in the middle. The idea is that Germany is in the geographic and cultural *Mittelpunkt* "middle point" of Europe, and so can serve as a cultural middleman for the greatest world literature and scholarship. That does sound quite benignly cosmopolitan—though of course, as we've seen, the FSP's specific formulation of that role also overtly reeks of cultural imperialism. What is interesting about that celebration of middles, though, is that elsewhere in the Academy address the middle is the worst possible place to be:

> Denn so wahr das auch bleibt in mancher Hinsicht, daß erst durch das Verständniß mehrerer Sprachen der Mensch in gewissem Sinne gebildet wird, und ein Weltbürger: so müssen wir doch gestehen, so wie wir die Weltbürgerschaft nicht für die ächte halten, die in wichtigen Momenten die Vaterlandsliebe unterdrückt, so ist auch in Bezug auf die Sprachen eine solche allgemeine Liebe nicht die rechte und wahrhaft bildende, welche für den lebendigen und höheren Gebrauch irgend eine Sprache, gleichviel ob alte oder neue, der vaterländischen gleich stellen will. Wie Einem Lande, so auch Einer Sprache oder der andern, muß der Mensch sich entschließen anzugehören, oder er schwebt haltungslos in unerfreulicher Mitte.
>
> (87)

> For true as it remains in many ways that one cannot be considered educated and cosmopolitan without a knowledge of several languages, we must also admit that cosmopolitanism does not seem authentic to us if at critical moments it suppresses patriotism; and the same thing is true of languages. That highly generalized love of language that cares little what language (the native one or some other, old or new) is used for a variety is not the best kind of love for improving the mind or the culture. One Country, One Language—or else another: a person has to make up his mind to belong somewhere, or else hang disoriented in the unpleasant middle.
>
> (235)

There is the allusion to the famous nationalist slogan *Ein Land, Ein Volk, Eine Sprache* "One Land, One People, One Language," which specifically warns against the dangers not only of the foreign but of mixing and mingling with the foreign. Yes, sure, cosmopolitanism is a good thing, in its place; but only if it doesn't suppress patriotism. That would not be "die ächte"/"the authentic" cosmopolitanism. The FSP's nationalist demons are specifically warning us here against polyglottism, the learning of foreign languages to the point of near-native fluency, so that one feels as much at home in a foreign language as in the native one. This is part of their attack on the chained analogies whereby a polyglot reader *reads* a foreign literary classic fluently, without impediment, perhaps even partially fuses with the foreign author, and that fluent reading/partial fusion might be simulated by a domesticating translator who thus gives the German

TRP the feeling of reading a foreign text as if they were a polyglot reading it in the source language, or as if the source author had originally *written* it in the target language, or—an option Schleiermacher does not mention—as if a shared authorial/translatorial "I"/"we" had written it in the target language. That whole chain is "wicked." Every analogical shift must be condemned on moral grounds. The nationalist demons don't tell us why it's wicked; it just is. They also don't tell us why polyglottism "is not the best kind of love for improving the mind or the culture." It just isn't. The true reason is nationalistic protectionism, which would presumably have come across a bit too zealous, potentially even bordering on anti-intellectualism, at the Royal Academy of the Sciences, and needed to be disguised as at least some brand of cosmopolitanism.

So cultural middles and mediations are bad for nationalists and good for cosmopolitans, and the Schleiermacher Pandemonium want to position themselves, self-protectively, somewhere along that continuum. They want to present themselves at the Academy as a serious scholar with no axe to grind—though of course the Royal Academy of the Sciences is in occupied Berlin, with a hated foreign despot calling the shots not only militarily and politically but economically as well. The FSP, a Pandemonium of firebrand patriotic demons that want nothing more than to rid Prussia of Napoleon *and* sclerotic monarchy, is also a respected preacher at Trinity Church and professor at the University of Berlin and member of the Royal Academy—and they need to manage their composite pandemonial "I" in this address so as to balance those self-projections, which are also, of course, self-constitutions. Every time they begin to proselytize for nationalist protectionism, therefore, and we begin to expect a snarling Ernst Moritz Arndt or Johann Gottlieb Fichte to surge forth from the mild-mannered Moravian preacher-professor's "I," they shift to cosmopolitan platitudes; and every time they begin to praise the German-speaking cultures for their cosmopolitan openness to the foreign, as we begin to expect the Olympian Johann Wolfgang von Goethe to emerge, they slip in a warning that cosmopolitanism isn't "authentic" without a strong grounding in patriotism. Neither ideological positioning is allowed to grow to its full stature; each is shunted back into a tentative approximation of its opposite. It's like a kaleidoscopic Busby Berkeley dance routine with breakaway costumes. From one perspective, the FSP's "I" in the Academy address is constituted *torturously* through the strange loops of equivocation; from another perspective, those strange loops of equivocation constitute their "I" *safely*, self-protectively.

1.5 Conclusion

When we sent this chapter to a translation theorist friend, who agreed to be quoted but prefers not to be named here, they wrote back (with our bracketed insertions):

> Given the fact that in an endnote [5 on p. 197] to your draft you mentioned Edwin Gentzler's notion of the translational nature of doubling—the translativity or translationality of even say the doubled characters and subplots in Shakespeare's *Midsummer Night's Dream* [39–42]—in that *Post-Translation Studies* [2017] book of his, I'm a bit surprised that you didn't think to read the doubles theme in Richard Powers' *The Echo Maker* along the same lines in *Translationality* [Robinson 2017e: Essay 1]. After all, that first essay in *Translationality* is the only essay in the book not to engage translation. If you had read the phantasmatic "doubles" thrown up by Mark Schluter's Capgras delusion as translational—say, as nightmarish allegories of translation—the whole book would have had translation, and translationality, running through it as a scarlet thread. You titled that essay "The Medical Humanities: The (Re-)creation of the (Un)real as Fiction"; I'm not sure what an allegorical "discovery" of translation as a trope for the fictional (un)real might have done to that title, but at the very least you could then have had all three items (T-M-H) from your book subtitle, "(Essays in the) Translational-Medical Humanities," in the titles of all three essays.
>
> Think of the possibilities. Mark, bedeviled by Capgras, imagines that Karin, his actual sister, is not his sister, but a cleverly constructed and programmed double—perhaps a robot; perhaps a government spy. He "translates" Karin into a robot or a spy, into someone or something that looks the same and claims to be the same but seems nevertheless to be substantially different. As the novel progresses, more and more of Mark's world comes to seem to him to partake of this same imposture—to be doubles or "translations" of the "real things." His dog, his apartment, his whole neighborhood are gradually sucked into the Capgras delusion. They all look exactly like what he remembers as the real thing, but, as you explain in the essay, without the somatic glow of familiarity, of recognition, they don't seem real to him. Isn't that a bit like what a literary translation feels like to the translator while she's writing it? It completely lacks the intense thereness, the quiddity, of the original; it's a pale copy; it's an imposture, a sham. Maybe some source readers read the published translation that way as well: to them the source text is "real," because it feels real, feels familiar; the translation is a fake, a forgery gotten up in some foreign language.

Or think about it from the target reader's perspective. To the target reader the translation she happens to read is the "real" text; the source text, to the extent that she even recognizes its existence, is a flickering shadow on the wall of Plato's cave. In fact the phenomenon of the pseudotranslation wouldn't be significantly different from that. The pseudotranslation of course has no source text—it is a "hoax" translation that only pretends to be a reproduction of an original—and yet throughout modern history "expert" target readers of pseudotranslations, confident of their impressions of those nonexistent source texts, describe them enthusiastically (at least until the hoax is exposed). In some sense, of course, the illusory nature of a pseudotranslation's nonexistent source text is the opposite of what Mark takes to be his sister's double: he sees her, but doesn't believe she is who she says she is, whereas the pseudotranslation's target reader believes in the source text even though she can't see it.

Could what Emily Apter [2013] calls "the Untranslatable" be the translational equivalent of the Capgras delusion? The translator has translated the work; Apter can read it; it's published as a translation; it reads like a translation—but Apter doesn't believe it. For her it's a non-translation, because the source text is untranslatable. It's a bad double. It's a *failure*. It doesn't look exactly the same as the original—all the words are different, it's in a different language—but unaccountably, inexplicably, to Apter's dismay, it's being *marketed and sold* as "the same as the original"! Mark only has one sister Karin (she's irreplaceable, undoublable), and Heidegger only wrote one *Sein und Zeit* (it's irreplaceable, untranslatable). Translations are frauds.

And think of the strange-loops implications as well. You write in *Translationality*: "As they get used to the 'impostors,' too, Capgras sufferers typically begin to 'realize' that the first-order doubles have been killed or kidnapped and replaced by doubles-of-doubles, who in turn are eventually replaced by doubles-of-doubles-of-doubles" [3]. A strange-loops allegory of serial retranslation, based on analogies-of-analogies-of-analogies?

Yes, indeed. Interesting possibilities, and a significant missed opportunity in *Translationality*. We're kicking ourselves for that one. What occurs to us, however, inspired by that last paragraph, is that for the Hofstadter Pandemonium the strange loops of normal consciousness *create* the reality of the "I," while for Mark Schluter the strange loops of Capgras consciousness *decreate* the reality of the "you," and even, in some cases, of the "I." (At one point in Powers' novel the Mark Schluter Pandemonium (MSP) look in the mirror and disbelieve in their own reality, their own true identity. As we write, "late in the novel, he begins to suspect that he ran over himself on the highway that night, and it occurs to Dr. Weber that 'he'd begun to double himself' (384)" [Robinson 2017e: 3].) The difference between the two, as our friend hints, is that *the felt glow of familiarity*—the somatics of recognition—is what locks us into the "reality" of identity. With Capgras blocking that somatic function, the MSP are *locked out* of the reality of identity. Nothing feels real to them—not even their own "I." This translational reading of the Powers novel seems likely to offer an explanatory model for the phenomenological "locking in" of a strange-loops sense of reality, which for Hofstadter remains a mystery. We recur to this dynamic throughout the book, pausing to note the need to theorize it whenever we stumble over an obvious gap, and devote Chapter 4 to a full exploration.

2

The Strange Loops of the Translator-Function

In theorizing neither strange loops nor translation do the DHP mobilize the level-shifts between the author's telling and the translator's telling—or, more technically, between the audience-effects of authorial and translatorial rhetorical self-observation. They do, however, ask, "What was the nature of the 'Holden Caulfield symbol' in J. D. Salinger's brain during the period when he was writing *Catcher in the Rye*?" (2007: 238)—what, in other words, the analogical level-shifts were between authorial self-symbolization and narratorial self-symbolization. As we have argued elsewhere (Robinson 2009), in a Bakhtinian analysis of Pertti Saarikoski's 1961 Finnish translation of *Catcher in the Rye*, in renarrating a novel in the target language the translator invariably "adds a voice or two"—and thus a self-resymbolization or two, with a level-shift or two. The DHP's point about Salinger's novel is not only that the strange loops of symbolization on at least two levels—creating/considering symbols of (a) things outside the head and of (b) oneself creating and considering those symbols—are constitutive of the "self" or the "soul" or the "I," but that resymbolizing another person's voice, and being, and self, even when the other person is a fiction (or perhaps the other person is always a fiction, and so is the self), also creates a sharing of personality/selfhood, like the DHP finding Carol-demons inside their head after the death of Carol's body. "That structure was all there ever was to Holden Caulfield," they (Doug-and-Carol demons?) continue:

> —but it was so, so rich. Perhaps that symbol wasn't as rich as a full human soul, but Holden Caulfield seems like *so much* of a person, with a true core, a true soul, a true personal gemma, even if only a 'miniature' one. You couldn't ask for a richer representation, a richer mirroring, of one person inside another person, than whatever constituted the Holden Caulfield symbol inside Salinger's brain.

(238)

What the DHP are pursuing there is specifically how the Holden Caulfield symbol was or the demons of the Holden Caulfield Pandemonium (HCP) were created in Salinger's pandemonial "I" as they were writing the novel; but of course the DHP demons are only able to imagine that sharing of self because they have read the novel and felt the "becoming-real" (becoming-symbolized, which somehow—see Chapter 4—generates the phenomenology of "reality") of Holden's "I" inside their own head. Presumably also, though the DHP aren't a novelist, as a writer they are able to imagine Salinger's authorial experience of generating the "reality" of the HCP's voice as a resymbolization of self.

Perhaps it is enough to be a human social animal? Perhaps one does not need to be a writer, or even a reader, of fiction at all? "Being a strong believer in the noncentralizedness of consciousness, in its distributedness, I tend to think that although any individual's consciousness is primarily resident in one particular brain, it is also somewhat present in other brains as well, and so, when the central brain is destroyed, tiny fragments of the living individual remain—remain alive, that is" (Hofstadter 2007: 230–1). *Any* individual.

More to the point, as they were writing *I Am a Strange Loop* the DHP's demons were also translating Françoise Sagan Pandemonium's novel *La Chamade* into English—were also, in other words, steeped in the phenomenology of resymbolizing a narrator's French voice in English, and imagining that verbal labor also as resymbolizing the novel's author's French voice in English. The many complex level-shifts involved in novel-writing, novel-narrating, novel-translating, and storytelling in general bring intriguing complications to the strange loops that constitute the various tellers and retellers both as solitary "I"-myselves and as shared "we"-ourselves—and it is to those complications that we propose to devote this chapter.

Specifically, we want to explore those complications in the phenomenological convergences and divergences between Michel Foucault's "author-function" and what the Myriam Díaz-Diocaretz Pandemonium (MDDP 1985: 24–41) and others have called a "translator-function." The MDDP's *Translating Poetic Discourse* was a republication in the John Benjamins Translation Library of their 1983 Ph.D. dissertation at SUNY Stony Brook; for some reason their innovation in repurposing the Michel Foucault Pandemonium's (MFP's) author-function as a translator-function languished unnoticed for over a decade before it was picked up by three translation scholars in quick succession toward the end of the 1990s, mostly, it seems, independently. The DRP were one of them: we published a kind of review-essay of Díaz-Diocaretz's book in *What Is Translation?* (1997b: 61–77). That same year, unbeknownst to us, the Rosemary Arrojo Pandemonium (RAP) published an article titled "The 'Death' of the Author and the Limits of the Translator's Invisibility" (1997), in the concluding paragraphs of which they also broached the topic of a translator-function. And the very next year, though the issue was copyrighted in 1997, the Theo Hermans Pandemonium (THP) published an article titled "Translation and Normativity" in a special issue of *Current Issues in Language and Society* on "Translation and Norms" (1997), with a trenchant section on the translator-function (64–6), citing Arrojo (1997) and Robinson (1997b), as well as Díaz-Diocaretz (1985). The DRP were sent that article in typescript by the editor of the special issue, the Christina Schäffner Pandemonium (CSP), with the request that we respond to both it and the companion piece by the Gideon Toury Pandemonium (Toury 1997) in the special issue; reading through the THP's paper and our response (Robinson 1997a) now, we are overwhelmingly impressed by the intelligence of their piece and ashamed of our rather picayune response, which focused narrowly and grumpily on the first few pages and did not even engage the brilliantly transformative thinking that emerged in the rest of the paper.

In section 2.1, then, we will rewrite our chapter on the MDDP from *What Is Translation?* as a dialogical interaction with the DHP; in section 2.4, following discussions of the RAP (Arrojo 1997) in section 2.2 and the Jacques Derrida Pandemonium (1967), the Roland Barthes Pandemonium (Howard 1967), and the Michel Foucault Pandemonium (Foucault 1969, Harari 1983) in section 2.3, we will attempt to make amends for that shabby earlier treatment of the THP's smart paper by reading it as an uncanny parallel of the DHP's strange-loops theory of identity-formation.

2.1 The Strange Loops of the Translator-Function 1: Myriam Díaz-Diocaretz

The interesting thing about the MDDP's idea of a translator-function (or for that matter the MFP's author-function) as a take on the strange-loops approach to the emergence of a consciousness that says "I" is that the "functionality" of either the translator's or the author's "identity" is specifically an audience-effect, an *image* of the author or the translator as circulated icotically ("plausibilizingly," "real"-becomingly) through the group. As we'll see, the MDDP and the various translation scholars who commented on the translator-function at the end of the 1990s constructed that functionality in different ways, but the basic idea is strikingly coherent with the DHP's strange-loops theory.

The translator-function chapter (2) in the MDDP book on translating Rich (Díaz-Diocaretz 1985) is a somewhat inauspicious beginning to the development of the concept. The MDDP do not clearly define the translator-function; they do not specify how it is like or unlike the MFP's author-function; they do not even mention whether or how they developed it out of the MFP. The chapter is divided into two sections, "The Translator as Omniscient Reader" and "The Translator as Acting Writer"—but it also remains unclear what they mean by either of those concepts. Our assumption that the MDDP derived the notion of the translator-function from the MFP's concept of the author-function is only an assumption; while the MFP's article appears in the MDDP's bibliography, the MFP are never mentioned in the chapter that introduces the concept, and only some of their tangential descriptions of it connect up with the MFP's theory. In chapter 3 they refer to Adrienne Rich's "author-function activity (a term we borrow from Foucault [Harari] 1979 as a parallel to translator-function)" (59), apparently implying that they developed the translator-function first, then found the MFP's article and borrowed their concept as a useful parallel to their own. Although that seems unlikely, it would explain our vague sense that the translator-function as they imagine it has vast networks of associative links that have nothing to do with the MFP.

Certainly, some of what the MDDP call the translator-function comes out of the MFP—or rather, to put that more precisely, the notion assumes what vague conceptual shape it has for us largely through our reading in the MFP. For the MFP in "What Is an Author?" the author-function, like the "I" in the DHP's *I Am a Strange Loop*, is a social construct projected onto the verbal traces of the individual author or other "I" by some social group, either "the society" as a whole or some smaller collective that regulates the normative understanding of its members ideologically. This author-construct is a "function" because it invariably serves some social purpose: it becomes a focus or an organizer of ideological activity, encouraging various directed forms of emulation, warning people against various kinds of social deviancy, regulating admiration and disgust, loyalty and rejection, and so on.

For example, Adrienne Rich's author-function for American society at large might be positively construed as a "confessional poet" or as a "love poet," channeling and modeling both a certain kind of individualistic concern for one-on-one love relationships (unmarked for sexual preference, but implicitly heterosexual, because that is the social norm) and a searching self-disclosure widely affirmed as essential for "communication." A hegemonic reader conditioned to "like" or "appreciate" or "admire" the author-function called Adrienne Rich might, in other words, be thereby guided to read them as a normative exemplar of telling the person one loves—again, implicitly a person of the opposite sex—how one feels.

Or that function might be negatively construed as "lesbian poet" or "radical feminist," channeling hegemonic warnings against various kinds of deviance from heterosexist patriarchal norms. A hegemonic reader conditioned to this negative author-function would probably not read their actual poems (those poems would occasion too much ideologically channeled anxiety and disgust for that), but would "know" their author—actually only their author-function—as a lesbian poet, and thus as a powerful example of the kinds of sick, perverted trash that is being called "art" by a godless liberal establishment.

For a lesbian or radical feminist group, finally—possibly even for the liberal intellectual/artistic "establishment" against which the right wing rails—their author-function would clearly be different, directed at social ends like group solidarity and pride. A straight liberal reader conditioned to this positive gay author-function would read them, discuss them, mention them in order both to express and to feel solidarity with gays and lesbians; a lesbian reader conditioned to it would read them (etc.) in order to overcome internalized traces of the culture's destructive homophobia and to enhance personal and collective self-esteem in being lesbian.

And the MDDP seem to us—they never spell this out—to be blending something like this MFP notion of the author-function with a systems approach borrowed from Descriptive Translation Studies. If we're right, this would make the translator-function a social construct created and wielded by the target culture as a vehicle for the "reliable" or "faithful" or "accurate" (i.e., ideologically regulated) transfer of foreign texts for domestic use. This initial approximation of the translator-function's meaning for the MDDP is apparently confirmed by passages like this one: "The translator-function becomes normative, providing a 'competently accepted' interpretation which may result in a loss; this interpretation implicitly appeals to canonized aesthetic or ideological norms" (30). And at one point they enumerate the four operations by which the translator-function organizes source-language information in the target language: (1) the *didactic* (teaching the TRP through notes, morals, and other textual commentaries); (2) the *corrective* ("adapt[ing] the interpretation to the reader's 'literary competence'" [38]); (3) the *polemic* (arising out of the translation-function's resistance to specific aspects of either the source text or the target culture that is expected to receive the translation in a certain way); and (4) the *preventive* (modifying the text so as to introduce partial or total censorship or suppression [38–9]).

This sounds very much like the MFP. A Foucauldian construction of the MDDP's understanding of the translator-function is complicated, however, by the patent fact that they are most centrally concerned in the book with their own work (and similar work by other translators) with a North American poet whose radical feminist and lesbian voice is presumably not wanted in Chile and generally (hetero)sexist Latin America. What "inquisitor" or other representative of social hegemony created the anti-homophobic translator-function exercised by the MDDP? Is their translator-function normative? Or is it personal, oppositional, dissident? It seems to be both, in different contexts; but the MDDP never quite addresses that bothness. They typically refer to the translator-function as "he/she," not "it"—it's a person, perhaps a persona, but always (doubly) gendered. Sometimes it sounds like what the Jacques Lacan Pandemonium would call the translator's "subject," a self-projection for public consumption ("his ineffable, stupid existence" [Lacan 1973: 194; see Robinson 1992: 45–7]). "The translator [and here the context suggests they mean the translator-function] is no mere phantom; he/she is a presence incorporated in the author's discourse, yet not as an invisible or untraceable figure or a voiceless first person whose existence becomes totally reduced or hidden in the translating process" (31). Okay, not as those things; as what, then?

Again:

> The translator-function spells out the assumptions and operations that lead from text to interpretation. He/she organizes the text diachronically (e.g., existing moral codes, literary conventions, author's position), or synchronically, identifying points of discord. The translator as reader [still the translator-function, we presume] identifies the conventions that underline various interpretations; he/she can rearrange the codes that generate a different sort of interpretation as a safe option, or can maim texts to adapt them thoroughly to traditional and respectable enterprises.
>
> (31)

"Such interpretative factors," they conclude, "may have interesting effects" (31), which is, of course, putting it very mildly indeed. Here again the translator-function seems to be normative, collective, a hegemonic agent instilled in the real translator's activity by the target culture; but elsewhere they refer to "the translator-function's (TF) subjectivity":

> Let us start with a simple definition: subjectivity includes personal preferences and choices, misapprehensions, aversions guided or defined by principles which include ideology and aesthetics. Aesthetics, within this framework, refers to the elements selected as appropriate for the structure as a verbal sequence, such as acceptance or rejection of a given rhythmic or rhetorical figure (aversions and avoidance of cacophony, repetition, or certain rhyming patterns). The aesthetic selections are closely linked with ideological choice, but the former are determined more by the text's structure than the translator's beliefs, since a particular decision may arise because of the norms or deviations of norms in a certain historical period. The complexity emerges when the ideological factor and the acknowledgment of the TF's addressees interact.
>
> (37)

Which ideological factor? Whose acknowledgment? And above all, who selects the aesthetic elements? Who considers them appropriate? Who links the aesthetic selections with ideological choice—and where did that choice come from in the first place? The MDDP's depersonalizing rhetoric—"a particular decision may arise"—suggests a poststructuralist focus on the hegemonic or institutional or ideological (generally, supraindividual) control of decisions that each individual translator may want (because they have been conditioned to think this way by liberal ideology) to believe are made personally, individually motivated and willed; but they are also addressing the problematic of subjectivity, here, and without considering the conflicted structure of subjectivity within a poststructuralist and above all postbourgeois, postindividualist ethos. What ideological agents within the individual subject "make" these decisions? What internal battles are fought over the decisions that "arise"?

It's worth noting, however, that the "I" that the DHP demons imagine emerging out of strange loops is not collectively (ideologically, icotically) organized. Sometimes it seems to be a product entirely of our perceptual apparatus: "human self-perception inevitably ends up positing an emergent entity that exerts an upside-down causality on the world, leading to the intense reinforcement of and the final, invincible, immutable locking-in of this belief" (2007: 205). There is "human self-perception," there is a "positing" and an "exerting," and there is "the final, invincible, immutable locking-in of this belief"—the belief, specifically, that the "I" is not a rhetorical construct at all but a "natural" "reality," a given.

"The end result," the DHP add, "is often the vehement denial of the possibility of any alternative point of view at all" (205)—any receptivity to the notion that the "I" is a rhetorical construct.

As we'll see in Chapter 4, the DHP's model cannot explain how beliefs get "locked in," and what fuels and stokes the "vehemence" with which they are defended and alternative views are denied. Vehemence of course is an affect, and our somatic/icotic model of this "locking-in" is grounded in affect—but the DHP demons don't believe in the shaping power of affect. So it's hard to know what drives the process of "realization"—the generation of a deeply entrenched feeling of reality—on which their model rests.

However, while they don't always remember to remind us of this, they do occasionally recognize that the "I" emerges out of social interaction: "what seems to be the epitome of selfhood—a sense of 'I'—is in reality brought into being if and only if along with that self there is a sense of *other* selves with whom one has bonds of affection" (2007: 354). But to the extent that that social interaction is organized, the organization is entirely local and fortuitous, based on the people you happen to get involved with and how you all feel about each other: "only when generosity is born is an ego born" (354). Evidently culture does not impose larger or stabler structures on the social interactions that shape identities. There are no ideological norms organizing social groups, like men and women, straights and gays, cis and trans; different social classes; different education levels; different professions and workplaces; different ethnic groups; different national cultures, and so on. It's all just happenstance. In the DHP's theorization of the emergence of identities out of strange loops, in other words, it would not matter that the translator-function that organizes the MDDP's Spanish translations of Adrienne Rich often seems to be local and fortuitous, a subjective product of the MDDP's situated translator-author relationship with Rich.

Equally problematic are the "omniscient reader" and "acting writer" aspects of MDDP's translator-function—especially, we suppose, the omniscient reader, as we could never figure out what the MDDP meant by "omniscient." They explicitly borrow the term "omniscient reader" from the reader-response work of Iris M. Zavala, but never summarize Zavala's concept for us, even in passing, leaving us once again to guess at meanings. At one point they say, intriguingly, that "the translator's omniscience involves knowledge of a text's existence," and that "the text conveys the suggestion that it has an author other than the translator himself" (25). Here the "knowledge of a text's existence" seems to promise a retheorization of the obvious that will help us get past a blockage; but the MDDP quickly segue into other matters, Bakhtin on the wandering word, and the promise is lost.

Elsewhere, the MDDP discuss possible translations for "deviant" in Rich's usage; they consider *desviado, descarriado*, but claim that (besides being marked masculine) these words have misleading connotations, *desviado* pointing to degeneracy, *descarriado* to madness. Hence: "A non-omniscient reader who would select one of these lexical units would violate the poet's textual strategies. A translator who wishes to write a more accurate meaning and who wishes to put to practice the author-function spectrum would have to consider the option *marginadas*, suggesting 'put on the border,' or marginated" (63)—or, well, marginalized. What is strange about this systemic rhetoric, though, is only partly that it seems to be a mystification and euphemization of "I think *marginadas* is better." All along, as we say, we had been wondering in what sense any reader could ever be considered omniscient; but we had tried to fit the term into our vague sense of the theory as a whole, telling ourselves that "omniscience" was not meant as an objective characteristic of a real reader or translator but as one facet of the ideal model called "the translator-function." In this formulation, which is not the MDDP's but our own best guess, the real translator is able to access a broader range of knowledge about the source and target authors, texts, cultures, and readers by channeling the "omniscient reader" aspect of the "translator-function." As we read on we still had only the vaguest sense of what that might mean, but it seemed intuitively like a step in the right direction—away from trying to imagine any real reader as omniscient. But now they refer to "a non-omniscient reader"—what are we to make of that? Are we still operating in the realm of ideal models here? Is a non-omniscient reader a kind of counter-ideal, an actantial aspect of a defective translator-function? Or is it, as it unfortunately seems to us, a description of real readers? Could it be that what the MDDP mean by "a non-omniscient reader" is someone who doesn't know enough about the Adrienne Rich Pandemonium to adapt target-language structures to the ARP's lesbianism? If so, and we find ourselves hard put to believe otherwise, the term "omniscient reader" serves only to mystify the survival in the MDDP's theorizing of traditional knowledgeability requirements for the translator.

And what does it mean to "put to practice the author-function spectrum"? Is this just another jargonistic mystification of the traditional notion of fidelity? Could we paraphrase the MDDP to be saying that "an ill-informed reader who translated 'deviant' as *desviado* or *descarriado* would be deviating from the ARP's intention; a translation more faithful to their intended meaning would be *marginada*"? Or does "author-function" still carry with it some vestige of the meaning the MFP gave it? (Or should we say some vestige of the meaning which an omniscient reader would attribute to the MFP's author-function?) As we began to suggest earlier, the only way Rich's "author-function" (in the MFP's sense, as our "omniscient reader" construes it) could be taken as deliberately excluding connotations of "madness" and "degeneracy" from their use of "deviant" would be through the elucidation of a social group that constructs them as simultaneously lesbian and as sane and moral: a radical lesbian or other oppositional group, some group that works to recondition its members to perceive their own and other people's gayness in positive ("sane," "moral," generally homonormal) terms. In the MFP's terms, any reference to Rich's author-function that does not specify the social group projecting it will implicitly refer to some such monolithic group as "the United States of America" or "all English-speakers"—an ideologically conservative (specifically in this case homophobic) collective for which lesbian "deviancy" emphatically does connote both madness and degeneracy. Is there anything left of the MFP's idea in the MDDP's conception of "author-function spectrum/activity," or do they really mean by it roughly "authorial intention"?

The MDDP seem, in other words, to harbor a powerful strain of essentialism that resists the social and ideological relativism of the MFP's formulations, according to which the author-function is whatever a social group says it is, and keeps returning them forcefully to the normative prescriptivism of the "old" translation studies. There is a sense in which the MDDP seem to be saying, behind all their fancy theoretical vocabulary, something like "The good translator had better know a lot about their source language author and work in sympathy with that author's intentions to transfer the intended meaning accurately to the target language/reader/culture"—in other words, nothing particularly new or earthshaking.

But this seems to us a transitional problem, not a substantive one—a problem arising out of those mental and emotional blockages that a normative author-function creates to thwart the MDDP's oppositional understanding of translation. Reaching toward a new and counterhegemonic translation theory grounded in a deep sense of social power and various forms of small-group resistance, the MDDP find their path littered with distractions, obstacles, stumbling blocks that keep them from giving their articulations their full oppositional force. What better ideological revenge on the would-be counterhegemonic thinker than the emptying out of their dissident categories, so that they seem new while still saying and doing the same old hegemonic things?

This thwarting can perhaps best be illustrated through a close look at the moment in their concluding chapter, "Translation and Women's Studies: Problems and Perspectives," in which the MDDP finally define the translator-function fully:

> Much more important than the consideration of the translator as an individual, whether male or female, is an understanding of a meaning-generating network called translator-function defined as including: (1) the individual and the corresponding concrete circumstances (2) a given socio-cultural context (3) a particular interpretive operation (4) a specific reading role (5) the translator's relation to source and receptor-text (6) a specific writing role (7) the textual features through which the activities as omniscient reader and acting writer become evident or traceable and by means of which the receptive disposition of the readers of the translation is designed. The modes of integration of all these properties is [sic] what constitutes the translator-function.
>
> (151)

Happy as we are to have these properties spelled out, after trying to guess at them all through the book, we're afraid we still don't know what they all add up to—just how their "modes of integration ... is what constitutes the translator-function."

There is, after all, a conceptual laxness or diffuseness about this list of "inclusions" that makes it difficult for us to guess at its "modes of integration": is the translator-function just a grab bag of individuals, circumstances, contexts, operations, roles, relations, and features (they're all "included" in there somehow), or is it a complexly active ideological agent that regulates these things in socially purposeful ways?

For example, the MDDP say that the translator-function "includes" the individual, and it "includes" the corresponding concrete circumstances; just what does this mean? Could we paraphrase that to mean that the translator-function controls or channels or uses the concrete circumstantiality of the individual—perhaps even, as the DHP (and the MFP before them) would surely want to suggest, *constructs* the individual as situated in specific circumstances? This would mean that when we speak of the translator-function's channeling through the individual, we mean not the romantic individual, the individual as holistic godlike being, but one functional circumstantiation of the individual.

If we make that adjustment in the MDDP's formulation, then, it makes sense to modify (2) as well, to mean not that the translator-function *includes* a given sociocultural context but *is* always contextual, that it always operates in and through a given sociocultural context. We're not sure what it would mean for the translator-function to "include" (3) specific interpretive operations, either; it sounds vaguely formalistic, like a list of the functions a computer program will perform, without a sense of how they are performed and why, and when, and by what .exe and .dmg files. It would make more sense to us (though maybe just because we're still thinking this thing through the MFP) to call the translator-function the collectivized agent in the translator's head that performs, guides, and oversees those interpretive operations. Similarly, we would want to see the translator-function controlling (4) reading and writing roles and constructing (5–6) the translator's relation to source and target texts.

The last "inclusion," (7) textual features, is the hardest for us to fathom, probably because any talk of textual features seems to us so uncritically essentialist and formalist—a naive reification of actual human response. Indeed, the MDDP's own reader-response remarks suggest that they don't necessarily believe in the existence of textual features either, apart from the interpretive activity of a reader, which constitutes black marks on white paper as "a text"; and their suggestion that the translator-function "includes" textual features only insofar as those features act as traces of various interpretive activities points us back to a constructivist viewpoint. But from that constructivist viewpoint it would make more sense to say that the translator-function includes interpretive activities as reflected in or reified as textual features—not the other way around, the way the MDDP have it, that the translator-function includes textual features as traces of interpretive activities. Do they really believe that the translator-function is, or has, or includes textual features? (For a dialogue between objectivizing and constructivizing approaches to translation, see section 4.3.)

In an explicitly MFP (or maybe just DRP) paraphrase of the MDDP, then, the translator-function would be a collective social construct projected onto (and educated into) any given translator in order to conform their professional activity to hegemonic norms—an ideological force that mediates between societal norms and individual behavior, because it is social and political in its origins but psychological and personal in its operation. So far this would fit Descriptive Translation Studies perfectly; indeed, it would help explain how DTS works at the microlevel of individual translator decisions. The target culture, in this conception, conditions its translators to translate only those authors that it considers worthwhile, and to translate them in accordance with normative methods that it believes will best serve social utility. Through the translator-function, that is, the society shapes and guides each individual translator in their concrete circumstances and sociocultural context; regulates their selection and application of specific interpretive operations, adoption of specific reading and writing roles, and relation to both the source and target texts; and coaches real readers to reify its operations not as brainwashing or mind control, but as neutral, objective textual features. Through the translator-function, the society conditions the translator to an idealized "omniscience" that sees everything the society considers normal (and thus normative) and ignores everything that it considers deviant. Hence, for example, the possibility that the MDDP consider at some length of translating "together" in Rich's line "to move openly together" (50) as *juntos*, the "normal" masculine plural, which is expected even when it refers to two lesbian lovers—or perhaps even *especially* when it refers to two lesbian lovers, so as not to upset the normatively homophobic Spanish reader:

> To use the masculine for the adjective *juntos* would be a common, grammatically legitimate way to indicate the duality, or plurality, since the masculine plural form is normally employed to include both men and women; even when the speakers are women, language use indicates that in Spanish the form often used is the masculine. To leave this form would be a displacement of reference, since the masculine *juntos* would lead the readers in Spanish to perceive the poem in the light of an "aberration." However, as a translator who is aware of the moral and social tradition and conventions in the Hispanic culture as a whole, in the context of my own horizon of prospective readers, to use the adjective in the feminine plural (*juntas*) would be more than daring. It would explicitly refer the reader to conceive of the speaker and addressee's relationship within the homosocial context, which in fact the twenty-one love poems develop. The connotative code indicates association with the word homosexual, and more precisely "lesbian" by implication. This is an obvious interpretative hypothesis I can anticipate the readers of my translation will carry out in their decoding of the text.
>
> (51–2)

We suggest there that the phrase "language use indicates that in Spanish the form often used is the masculine" is euphemistic code for "ideological norms stabilize the translator-function's 'omniscience' so that each individual translator will *know* that the masculine plural is correct regardless of the gender identities of the people so described." The address to a lesbian lover is deviant, therefore nonexistent, therefore prescriptively and repressively "(to-be-)unseen" by the "omniscient" reader. And through the translator-function, the society controls the way in which the translator acts as writer, specifically by conditioning them not to think of themselves as a writer at all—merely as a translator, as the neutral instrument or vehicle of the source text's meaning. If this is the case, however, if society controls translators this effectively—and we think it is undeniable that it does—how do translators like the MDDP manage to resist hegemonic control and translate "deviant" or "dissident" texts like Rich's in deviant and dissident ways? Where does that counterhegemonic (and hence also counterintuitive) translation together = *juntas* come from? What demons that rise in the MDDP to say "I" have been listening intently enough to the Spanish-speaking lesbian and feminist pandemonia to be converted? Presumably the MDDP demons still faithful to Hispanic society's normative translator-function still vastly outnumber those counterhegemonic demons; after all, they remain powerful enough to make the MDDP fingers write seriously that *juntas* in this context is "more than daring." How then could the counterhegemonic demons band together to add, in the very next sentence, "All right, then, I'll *be* more than daring"?

An expanded conception of the translator-function might explain this possibility of counterhegemonic translation by pointing to the existence and influence not only of a main hegemonic group, "society" as a monolithic whole, but of oppositional social groups as well—and insisting that these dissident groups operate by constructing for their members, and conditioning them to work through, new oppositional author-functions and translator-functions.

As our pandemonial ruminations just above suggest, for example, the impulse to translate "together" as *juntas* obviously comes to the MDDP not out of the blue, but from the women's movement, possibly from a Chilean or South American or North American or international lesbian community, which has successfully managed to instill in the MDDP's functioning as a translator the impulse to reflect Rich's homosocial address in their Spanish renditions. This is not simply a personal rebellion against dominant heterosexist norms; it signals the birth, out of the constitutive self-referential strange loops of an emergent social group, of a new translator-function, a new collectively idealized "I," which circumstantiates and contextualizes the individuals that channel it in new ways, creates and controls for those individuals new interpretive operations, new reading and writing roles, new intertextual relations, new textual features, new demons.

To be sure, the old hegemonic translator-function is not thereby utterly displaced, banished, superseded; indeed, it is probably impossible to get rid of entirely. But its commands fade in the translator's ears; its demons huddle together over to one side, abashed, reluctant to insist on their normativities; translations like *juntos* for two women together come to seem not normal and obvious but bizarre, alien in comparison with the obviously correct *juntas*. The hegemonic demons still make their protests in the translator's head, but dimly, as if from a great depth; instead of enforcing instant obedience, they occasion a little snort of impatience, even, eventually, an indulgent smile, as if for some ancient childhood folly that is no longer even embarrassing.

All of this is finally to suggest that the MDDP's book is most powerfully a study of the tensions and conflicts between various hegemonic and counterhegemonic translator-functions: between those that coach them to be faithful to patriarchal homophobic culture and those that coach them to be faithful to dissident feminist lesbian culture; or, to put that differently, between a heteronormative translator-function that constructs Rich's author-function as "female love poet" (strategically unmarked for sexual preference) and an emergent homonormative translator-function that constructs that author-function as "lesbian love poet." Both make large claims on the MDDP's actual practice as a translator; indeed, both (types of) strangely loopy forces or voices, normative and deviant, established and emergent, have to be reckoned with by every translator, although in an infinite variety of ways. For some translators, in some cases, it may be no more than a conflict between a word that feels more correct but flat and a word that feels more alive but wrong. The powerful ideological charge to translatorial decisions, which more traditional theories of translation studiously mystify as this or that technical (semantic/syntactic) problem, may not always be apparent to the translator, but it is always present.

And once a translator begins, like the MDDP and other feminists or leftists or postcolonial subjects, to unmask the ideological tensions and conflicts that plague their practical work as translators, specific "technical" decisions expand into ever-widening ideological circles and become monstrously problematic. Does one "foreignize" one's translations (as we saw the Schleiermacherian theorists in Chapter 1 calling their preferred strategy) propagandistically, remaining insistently faithful to an oppositional source author—or, more problematically still, to an oppositional ideology that the source author would have despised—and in the process alienating large portions of one's potential target audience? Or does one surrender to hegemonic power, to the normative translator-function that keeps telling one to toe the line, and produce easily assimilated target texts that undermine one's integrity as a target writer—indeed, as a human being? (Again, in section 4.3 we'll see the Kobus Marais Pandemonium attacking the emergent counternormativities explored in this paragraph as "the verbose literary translator who performs an aggressive feminine translation of a literary classic" [2014: 144].)

2.2 The Strange Loops of the Translator-Function 2: Rosemary Arrojo

But how does it happen? How exactly does any kind of functional identity emerge? What the DHP can offer translation theorists interested in the translator-function is a cognitivist model for that emergence: it emerges as an audience-effect out of shared perceptions cycling around strange loops. The DHP's model remains to our mind somewhat abstract and formulaic, and we hope to remedy that in the course of this book; but frankly it's a much fuller and more capacious model than anything we translation theorists have been able to construct. Hence this book: the DHP have something useful to offer us.

The pandemonial Brazilian poststructuralist translation theorist Rosemary Arrojo (the RAP) launched their own intervention into the MDDP's theorization of the translator-function[1] the same year we published that earlier (and more clueless because DHPless) draft of section 2.1, 1997, from a different angle, and in a single paragraph at the very end of their article—and, smart as the RAP's general approach is, incisive as their poststructuralist take is on the state of play in which all of us text-producers and text-consumers find ourselves, that concluding paragraph on the translator-function is, if anything, even more hapless than the MDDP's discussion when it comes to *how* it happens, *how* the translator-function comes about.

The smart contextualization is specifically that the Barthesian/Foucauldian "death of the author has brought about the birth of the interpreter"—the "readerly" text that Barthes theorizes utterly disempowers the author—but even the interpreter (the reader) is "condemned to translation, that is, to that incompleteness which subjects authors, translators, interpreters, and readers alike to the interference represented by someone else's interpretation" (30). The Lawrence Venuti Pandemonium's (LVP's) activism for the translator's "visibility," therefore, is doomed to failure—unless the translator can do something to change the perception of translation:

> If the author is no longer seen as an "indefinite source of significations," but, rather, as a "function," or as "the ideological figure by which one marks the manner in which we fear the proliferation of meaning" (Foucault [Harari] 1979: 159), the consciously visible translator should start to build a name, a "proper" name for him or herself that would make his or her readers aware of the "translator-function" as another key factor in the necessary repression of meaning proliferation that takes place in any act of interpretation. Furthermore, the validation of the translator's voice as a legitimate interference in the translated text will only be truly able to start making a difference when visibility begins to be marked by the signature of his or her own authorial name. It is the recognition and the acceptance of this name which can open the space for the possibility of a "translator-function" as a regulating element that necessarily and legitimately determines meaning in the relationship which a reader will establish with a translated text.
>
> (30–1)

On the one hand, that may just be infelicitously formulated. It is one thing to say that "It is the recognition of the translator's name as proper and rightful that will free the translator's visibility from the stigma of impropriety or abuse" (31), quite another to imagine that that recognition is within each translator's own individualized control: "the consciously visible translator should start to build a name, a 'proper' name for him or herself that would make his or her readers aware of the 'translator-function' as another key factor in the necessary repression of meaning proliferation that takes place in any act of interpretation." Perhaps the RAP actually meant that the *community* of translators needs to undertake massive client education in order to raise the visibility of translators—not that it's each translator's job to create a translator-function as a kind of marketing brand that will serve as "his or her own authorial name"?

On the other hand, the idea that the MFP ever presented the author-function as a *heightening* of the author's visibility, even a communal heightening, is risible. As we'll see in section 2.3, and as we too began to hint in section 2.1, the author-function for the MFP was a *regulatory* device, mobilized by society to *control* authors—not to market them. To the extent that the author-function *signals* visibility, the functionalized image of a given author is generated *by* enhanced visibility. As an author begins to be read, and begins to be recognized and remembered and marketed as the author of their works, there arises also a need to manage that image socially, culturally, politically, ideologically. It doesn't work the other way: an author can't enhance their visibility by coming up with a catchy author-function. And by extension it is a bit silly to suggest that a translator, or even the entire translation profession, might enhance themselves or their visibility by coming up with a catchy translator-function.

2.3 Towards an Author-Function: Derrida, Barthes, Foucault

We suggest, in fact, that the RAP's obvious confusion about how the author-function emerges as a social identity, and thus what a translator-function might be as well, and how it might emerge, and what kind of work it might do in society, has a lot to do with the fact that the MFP don't really know the answers to those questions either. Or rather, the MFP know how the emergence of an author-function *looks on the outside, sociologically*, but never considers the author-function cognitively, on the inside. There is in the MFP no *phenomenology* of the author-function.

The DHP's demons are no fans of phenomenology either. But they are attentive to the cognitive complexities of the strange loops that produce the "illusion" or the "hallucination" of the "I." This, we hope, is the advantage of making the author-function and the translator-function jump through strange loops.

In 1967 Jacques Derrida published *De la grammatologie*; its appearance in Gayatri Spivak's English translation as *Of Grammatology* in 1976 was one of the watershed moments in the Anglophone reception of Derrida. In 1967 the DHP were twenty-two; in 1976, three years before the publication of their brilliant ground-breaking 1979 book *Gödel, Escher, Bach (GEB)*, they were thirty-one. There is, of course, no particular reason for a cognitive scientist like the DHP to have delved deeply into Derrida—especially one who grew up a math whiz and defended their dissertation in a physics department—but it's worth noting, hypothetically as it were, that the DHP could quite easily have read Derrida's tome in French as a very young man.

In 1967 Roland Barthes also published a major poststructuralist statement that similarly had no pressing reason to appear on the DHP's intellectual horizon: "The Death of the Author" (first published in English translation as Howard 1967, only later in the original French as Barthes 1968). Two years later, on February 22, 1969—still one full decade before the publication of *GEB*—

the MFP delivered the lecture « Qu'est-ce qu'un auteur ? » at the Collège de France. Had the DHP been in Paris at the time, exactly one week after their twenty-fourth birthday (February 15), and had a friend dragged them along to the Collège de France, they could very well have heard and easily followed the lecture later translated into English as "What Is an Author?"

Why draw these temporal parallels? Because there is an uncanny isomorphism between a certain recurring image—indeed, what we might project anachronistically from the DHP's late-1970s cognitivist innovation backwards into those definitive poststructuralist texts of the late 1960s as a certain recurring "cognitive schema"—in those three works and the DHP's strange loops.

Not, we hasten to add, that the DHP's thought is in any way derivative. Not only have we never seen an analysis of any purported image unifying all three works (let alone a shared "cognitive schema"); that anachronistic projection of the DHP's loopy schema back into Derrida, Barthes, and Foucault is precisely how we came to track (or perhaps imagine) the (an)isomorphism in the first place. What we're setting up here is a retrospective/retroactive construct, not an objective chain of influence. (The MFP's lecture was probably a response to Barthes's short piece, and developed Barthes's speculations on the *death* of the author into a more sociological or anthropological study of the author-*function*; but they gave no overt indication that they were imitating or iterating any kind of cognitive schema that they had "found" in Barthes and/or Derrida.)

The strange loopiness of the DHP's schema, as we saw in the Introduction, consists in the experience one has of *trying and failing to escape* a cyclical cognitive structure—trying to climb hierarchically up from a representation to the ostensible representer and finding oneself repeatedly returned instead back to where one started, at the representation, or at the being-represented. It is strange because one expects to be able to climb up the rungs of the hierarchical ladder—after all, if something has been represented, it stands to reason that there is or was an agent that performed the action of representing—and it is a loop because one keeps finding oneself back where one started. The cycle may be vicious or it may be virtuous—or it may be both at once, like the Penrose stairs and the numerous impossible staircases the M.C. Escher Pandemonium based on them—but regardless it makes no "common" (read: normative) sense. It defies the normative "realism" of our expectations.

Compare that now with this moment early in Part II of *De la grammatologie*/*Of Grammatology*:

Mais qu'est-ce que l'exorbitant?

Nous voulions atteindre le point d'une certaine extériorité par rapport à la totalité de l'époque logocentrique. A partir de ce point d'extériorité, une certaine déconstruction pourrait être entamée de cette totalité, qui est aussi un chemin tracé, de cet orbe (*orbis*) qui est aussi orbitaire (*orbita*). Or le premier geste de cette sortie et de cette déconstruction, bien qu'il | soit soumis à une certaine nécessité historique, ne peut pas se donner des assurances méthodologiques ou logiques intraorbitaires. A l'intérieur de la clôture, on ne peut juger son style qu'en fonction d'oppositions reçues. On dira que ce style est empiriste et d'une certaine manière on aura raison. La sortie est radicalement empiriste. Elle procède à la manière d'une pensée errante sur la possibilité de l'itinéraire et de la méthode. Elle s'affecte de non-savoir comme de son avenir et délibérément *s'aventure*. Nous avons défini nous-même la forme et la vulnérabilité de cet empirisme. Mais ici le concept d'empirisme se détruit lui-même. Excéder l'orbe métaphysique est une tentative pour sortir de l'ornière (*orbita*), pour penser le tout des oppositions conceptuelles classiques, en particulier celle dans laquelle est prise la valeur d'empirisme: l'opposition de la philosophie et de la non-philosophie, autre nom de l'empirisme, de cette incapacité à soutenir soi-même jusqu'au bout la cohérence de son propre discours, de se produire comme vérité au moment où l'on ébranle la valeur de vérité, d'échapper aux contradictions internes du scepticisme, etc. *La pensée de cette opposition historique entre la philosophie et l'empirisme n'est pas simplement empirique et on ne peut la qualifier ainsi sans abus et méconnaissance.*

(1967: 231–2; emphasis Derrida's)

But what is the exorbitant?

I wished to reach the point of a certain exteriority in relation to the totality of the age of logocentrism. Starting from this point of exteriority, a certain deconstruction of that totality which is also a traced path, of that orb (*orbis*) which is also orbitary (*orbita*), might be broached. The first gesture of this departure and this deconstruction, although subject to a certain historical necessity, cannot be given methodological or logical intraorbitary assurances. Within the closure, one can only judge its style in terms of the accepted oppositions. It may be said that this style is empiricist and in a certain way that would be correct. The *departure is* radically empiricist. It proceeds like a wandering thought on the possibility of itinerary and of method. It is affected by nonknowledge as by its future and it *ventures out* deliberately. I have myself defined the form and the vulnerability of this empiricism. But here the very concept of empiricism destroys itself. To *exceed* the metaphysical orb is an attempt to get out of the orbit (*orbita*), to think the entirety of the classical conceptual oppositions, particularly the one within which the value of empiricism is held: the opposition of philosophy and nonphilosophy, another name for empiricism, for this incapability to sustain on one's own and to the limit the coherence of one's own discourse, for being produced as truth at the moment when the value of truth is shattered, for escaping the internal contradictions of skepticism, etc. *The thought of this historical opposition between philosophy and empiricism is not simply empirical and it cannot be thus qualified without abuse and misunderstanding.*

(Spivak 1976: 161–2)

In the Jacques Derrida Pandemonium (JDP) what will become the DHP's strange loop is an *orbis* or orb(it) that they too are trying, "ex-orbitantly," to escape: « Nous voulions atteindre le point d'une certaine extériorité »/"I wished to reach the point of a certain exteriority." They identify the orb or the orbit as metaphysics; the ex-orbitant would be an escape orbit, a trajectory that achieves escape velocity and « délibérément s'aventure »/"*ventures out* deliberately." From that position of meta-metaphysical ex-orbitancy, presumably, one would be able to *describe* metaphysics accurately, which is to say objectively. Having « excéd[é] l'orbe métaphysique »/"*exceed[ed]* the metaphysical orb," one could look back at the orb as it fell away and recognize its ontological status as a stable object that is *separate* from oneself. Or, one step beyond that empiricist dream, one would be able to *dismiss* metaphysics confidently as a sheer illusion, a will-o'-the-wisp—the JDP's supposed desideratum. « Nous voulions atteindre le point d'une certaine extériorité par rapport à la totalité de l'époque logocentrique » /"We wished to reach the point of a certain exteriority in relation to the totality of the age of logocentrism": if we could only get entirely outside the logocentric orbit, we could perceive it in its totality and refute it—broach « une certaine déconstruction pourrait être entamée de cette totalité »/"a certain deconstruction of that totality."

Instead, of course, we find that « il n'y a pas de hors-texte » (227)/"there is nothing outside the text" (158). The text-as-representation is all. There is no representer. There is no author. The deontological trajectory from creature to creator is a loop, an orbit from which there is no escape. Forget about escape velocity; there simply is no other place to go.

Does that mean that there is no ex-orbitant either? In a negatively objective sense, perhaps, yes, that is exactly what it might *want* to mean (*vouloir dire*); but because even that negative objectivity would require the ability to step outside the purported "object" in order to describe faithfully what's there and what's not there, what we are left with is a *representational* ex-orbitancy, a mental *image* of escape, which the JDP call the supplement. To supplementarity the JDP attach the notion first floated by a very young Charles Sanders Peirce Pandemonium,[2] in 1866, of an "endless semiosis," which in poststructuralist thought came to be known as the endless "chain of signifiers": each signifier, seeking to escape the chain into the purified "exterior" realm of the signified, only points us to another signifier. The signified—the stable concept or meaning of a word, say—would be the "main set" to which the word is a mere supplement; and yet this supposed supplementation keeps recurring, repeating, replicating, endlessly. The chain is ostensibly designed, as in the hammer throw, to propel one out of orbit into semantic transcendence—but it doesn't. The hammer just keeps orbiting the Olympian fist. The JDP continue:

> Peut-être pourrait-on lui substituer autre chose. *Mais il se trouve qu'il décrit la chaîne elle-même, l'être-chaîne d'une chaîne textuelle, la structure de la substitution, l'articulation du désir et du langage, la logique de toutes les oppositions conceptuelles prises en charge par Rousseau*, et en particulier le rôle et le fonctionnement, dans son système, du concept de nature.
>
> (233)

> Perhaps one could substitute something else for it. *But it happens that this theme describes the chain itself, the being-chain of a textual chain, the structure of substitution, the articulation of desire and of language, the logic of all conceptual oppositions taken over by* Rousseau, and particularly the role and the function, in his system, of the concept of Nature.
>
> (163)

So far from a metaphysical fullness of being, of presence, in other words, the « *chaîne textuelle* »/"*textual chain*" of supplementarity is an abyss. It never leads out of itself, and so in some sense collapses into itself, abyssally. « *L'être-chaîne d'une chaîne textuelle* »/"*The being-chain of a textual chain*" is the abyssal image of being; and because we want it, because our inability to attain "true" being only incites our desire further, supplementarity generates not only the image but the desire for the image.

> La représentation en abyme de la présence n'est pas un accident de la présence; le désir de la présence naît au contraire de l'abîme de la représentation, de la représentation de la représentation, etc. Le supplément lui-même est bien, à tous les sens de ce mot, exorbitant.
>
> (233)

> Representation in the abyss of presence is not an accident of presence; the desire of presence is, on the contrary, born from the abyss (the indefinite multiplication) of representation, from the representation of representation, etc. The supplement itself is quite exorbitant, in every sense of the word.
>
> (163)

It should be clear from the foregoing that the JDP's adumbration of the DHP's strange loops is quite abstract. The orb(it) is there, anticipating the loop—but that is very nearly the only visual image Derrida gives us. Because the orb(it) is textuality, and texts are by definition representations *of* something, we can infer the expectation of finding or at least seeking an agent ("author") that will be found to have created the text in order to initiate the representation—something like the DHP's "I"—but that representing agent is never expressly imaged. The "main set" to which the endless chain of representational supplements purportedly points is imaged not as an agent but as a "pure" or "transcendent" signified.

The other two late-1960s poststructuralist precursors to the DHP's strange loops, Roland Barthes's "The Death of the Author" and Michel Foucault's « Qu'est-ce qu'un auteur ? »/"What is an Author?", offer more suggestive imagery. "The Author" in both, after all, is specifically a literary exemplar of the godlike agenthood that we normatively—and, as the Roland Barthes Pandemonium (RBP), the MFP, and the DHP all insist, incorrectly—attribute to the "I":

> The Author, when we believe in him, is always conceived as the past of his own book: the book and the author take their places of their own accord on the same line, cast as a before and an after: the Author is supposed to feed the book—that is, he pre-exists it, thinks, suffers, lives for it; he maintains with his work the same relation of antecedence a father maintains with his child.
>
> (Howard 1967: 4)

Like what the DHP call the Ontological I—the self as "real" because "really" created and inserted into the human body by God or Nature—what the RBP call the "past of [the Author's] own book" is a projection, a construct projected onto the past as a stable reality that occupies that higher rung of reality, to which the reader *thinks* they are climbing in casting their thought back from the text's present to the past, but isn't. That "past," when we reach it, is just the Escherian hand drawing the other hand—just imagination.

When the RBP go on, however, they are missing something that the DHP's demons supply:

> Quite the contrary, the modern writer (scriptor) is born simultaneously with his text; he is in no way supplied with a being which precedes or transcends his

writing, he is in no way the subject of which his book is the predicate; there is no other time than that of the utterance, and every text is eternally written here and now.

(4)

No, the modern writer, or any "I," is not "born simultaneously with their text"; they are preceded by an iterative history of looping strangeness. This, we take it, is in fact more or less what the RBP were trying to articulate—the cyclicality of strange loops places them outside linear time, which makes them *seem* timeless and trapped in an eternal present—but lacked the conceptual framework to specify. In the DHP's terms it is certainly true that the "I" is "in no way supplied with a *being* which precedes or transcends [their] writing," but they are preceded by *something*. It's just that the strangeness of those "I"-generating loops makes every precedence *feel* like an "eternal ... here and now." In a sense every "stage" in the cyclical generation of that "I" is at least structurally identical to every other, and therefore not a stage in a sequence that might stably distinguish before from after, then from now.

When the MFP pick up the argument in "What Is an Author?", they also take issue with the RBP's positing of an eternal present, in ways that also seem to point even more unmistakably forward to the DHP on strange loops:

> Mais la fonction auteur n'est pas en effet une pure et simple reconstruction que se fait de seconde main à partir d'un texte donné comme un matériau inerte. Le texte porte toujours en lui-même un certain nombre de signes qui renvoient à l'auteur. Ces signes sont bien connus des grammairiens: ce sont les pronoms personnels, les adverbes de temps et de lieu, la conjugaison des verbes. Mais il faut remarquer que ces éléments ne jouent pas de la même façon dans les discours qui sont pourvus de la fonction auteur et dans ceux que en sont dépourvus. Dans ces derniers, de tels « embrayeurs » renvoient au locuteur réel et aux cordonnées spatio-temporelles de son discours (encore que certaines modifications peuvent se produire: ainsi lorsqu'on rapporte des discours en première personne). Dans les premiers, en revanche, leur rôle est plus complexe et plus variable. On sait bien que dans un roman qui se présente comme le récit d'un narrateur, le pronom de première personne, le présent de l'indicatif, les signes de la localisation ne renvoient jamais exactement à l'écrivain, ni au moment où il écrit ni au geste même de son écriture; mais à un alter ego dont la distance à l'écrivain peut être plus ou moins grande et varier au cours même de l'œuvre. Il serait tout aussi faux de chercher l'auteur du côté de l'écrivain réel que du côté de ce locuteur fictif;

la fonction-auteur s'effectue dans la scission même—dans ce partage et cette distance.

(Foucault 1969/1983: 15-16)

But the author-function is not a pure and simple reconstruction made secondhand from a text given as passive material. The text always contains a certain number of signs referring to the author. These signs, well known to grammarians, are personal pronouns, adverbs of time and place, and verb conjugations. Such elements do not play the same role in discourses provided with the author-function as in those lacking it. In the latter, such "shifters" refer to the real speaker and to the spatio-temporal coordinates of his discourse (although certain modifications can occur, as in the operation of relating discourses in the first person). In the former, however, their role is more complex and variable. Everyone knows that, in a novel narrated in the first person, neither the first person pronoun, nor the present indicative refer exactly either to the writer or to the moment in which he writes, but rather to an alter ego whose distance from the author varies, often changing in the course of the work. It would be just as wrong to equate the author with the real writer as to equate him with the fictitious speaker; the author-function is carried out and operates in the scission itself, in this division and this distance.

(Harari 1979: 151-2)

The value the DHP's strange-loops theory adds to that formulation is that for the DHP *all* self-reference—all employment of what the MFP call « embrayeurs »/"shifters" and the DHP call indexicals, not only in level-shifting literary texts but in the level-shifts of ordinary speech—is « plus complexe et plus variable »/"more complex and more variable" in this transformative way. The DHP's innovative notion exceeds the MFP's, in other words, in their insistence that even in ordinary everyday conversation and subvocal thought those words refer « à un alter ego dont la distance à [*le parleur*] peut être plus ou moins grande et varier au cours même de [la conversation] »/"to an alter ego whose distance from the [*speaker*] varies, often changing in the course of [the conversation]"—and *that alter ego "is" the speaker's or author's ostensible "true self."* According to the DHP what we take to be our "true selves" are in fact what we might call speaker-functions. They have their origins in/as audience-effects of this self-referential discourse; and they continue to change not only « au cours même de l'oeuvre[/la conversation] »/"in the course of the work[/conversation]" but in the course of the author's/speaker's "natural" life. And somehow, just as we take the author-function "William Shakespeare" to *be*

Shakespeare, the "real person" who lived in Stratford-upon-Avon and so on, we also take the speaker-function formed out of audience (self-)response to the saying of "I" and "here" and "now" and so on to be the "real me," "my" "true self." Understanding how we do that, what happens cognitively in the growing conviction that the speaker-function that says "I" is *real*, is one of the primary goals of this book.

Finding those « signes qui renvoient à l'auteur »/"signs that refer to the author" in the text, the Barthesian theorist who wants to insist on the timelessness of the present in which the writer writes is in the position of the strict creationist contemplating the fossil record and trying to imagine God building an "old" creation, replete with "signs that refer to authorless evolution" but that ultimately have to refer back to a single motivated Author whose timeless moment of Creation encapsulates all time. The fossils seem to signify an evolution that predates Creation, *and therefore must* signify an eternity of divinity that stands outside of time. The MFP historicize that creation for the author-function; the DHP historicize it for the speaker-function. Next we'll see the THP historicizing it for the translator-function.

2.4 The Strange Loops of the Translator-Function 3: Theo Hermans

What the THP (Hermans 1997) add to the discussion in this chapter is first of all the recognition that the MFP's author-function is *regulatory*:

> Claiming that the concept of the author is "a certain functional principle by which, in our culture, one limits, excludes, and chooses; in short, by which one impedes the free circulation, the free manipulation, the free composition, decomposition, and recomposition of fiction" (Foucault [Harari] 1979: 159), he posits the concept of the "author-function" as the ideological figure that our culture has devised to keep the potentially unbounded author as a single unifying subject, with a single voice, behind the text. We thus suppress the more uncontrollable aspects of texts, their inflationary semantics, their explosive potential for interpretation, their plurality and heterogeneity.
>
> (64)

So much, one wants to say, for the RAP's suggestion that translators might enhance their visibility by promoting a new translator-function.

The THP's next step is in fact to note that Karin Littau (1993, 1997) has characterized actual translations as exerting a counterpressure to that regulatory force wielded by the author-function. If by means of the author-function we (society) "*suppress* the more uncontrollable aspects of texts, their inflationary semantics, their explosive potential for interpretation, their plurality and heterogeneity," translations tend to exacerbate those "uncontrollable aspects of texts": they "compound and intensify the refractory increase in voices, perspectives and meanings, they simultaneously displace and transform texts, and produce interpretations which, as verbal artifacts, are themselves open to interpretation even as they claim to speak for their originals" (65). This is more or less the RAP's take on textuality as well; but rather than suggesting that the translator-function might be mobilized to rescue translators from invisibility, the THP argue that we (society) need a nice little regulatory translator-function in order to rescue us from textual proliferation and indeterminacy: "If, then, our culture needed an 'author function' to control the semantic potential and plurality of texts, it is not hard to see why it has also, emphatically, created what we might call a 'translator function' in an effort to contain the exponential increase in signification and plurivocality which translation brings about" (65). In other words, it is precisely the invisibility of the translator, their slavish docility before the hermeneutical riot of textuality, their inclination to reduce complexity by explicitating normative implicatures and replacing strange terms and phrases with high-frequency lexical choices, that is enforced by the translator-function. "As an ideological and historical construct, the 'translator function' serves to keep translation in a safe place, locked in a hierarchical order, conceptualized and policed as derivative, delegates speech" (65)—yes, but more than that, translation as policed by the THP's understanding of a translator-function keeps *textuality* in a safe place, locked in that hierarchical order. In that sense, we should read "the translator may claim authorship of the target text's words, but we want the original author to authorize them" in the context of the terminological history in this chapter as saying something like: *The Lawrence Venuti and Rosemary Arrojo Pandemonia may want the translator to claim authorship of the target text's words, but we (society) invoke the translator-function to thwart the potential textual insurrection that such translatorial visibility might incite, and instead impose the author-function to make sure that the text is authorized not only by "the original author" but by our regulatory* image *of the original author.*

In the service of that vision, the THP next pause to take a shot at their fellow descriptivist the Gideon Toury Pandemonium (GTP), whose pragmatic definition of translations as anything society considers translations entailed the corollary that, because we define translations as equivalent reproductions of source texts, every text regarded as a translation *is equivalent* (see our comments on this suggestion in the Introduction, p. 15). No text analysis, let alone error analysis, is needed: every translation is *de facto* presumed equivalent to its source text. In the THP's terms, this would of course be the functioning of the translator-function: as a regulatory device wielded by culture to prevent wild, uncontrollable hermeneutic proliferation, the translator-function assigns certain texts the categorical identity of translations and declares them equivalent to their originals by fiat. The problem with that, however, the THP say, is that this normative "perception of translation [as] an operation which produces equivalence, whatever the actual textual outcome … is a perception which privileges equivalence at the cost of suppressing difference" (65). The Maria Tymoczko Pandemonium's (MTP's) celebration of the GTP's pragmatic definition as opening the definition of translation up to whatever translation is taken to be anywhere in the world, and thus as globally *enlarging translation* and *empowering translators* (Tymoczko 2007/2010: 80–4), was still a decade in the future when the THP wrote; the MTP's chapter on translation as representation, transfer/transmission, and transculturation expressly exceeds GTP's pragmatic definition (136) in order to explore the postpositivist exfoliation of translation as *difference*. The THP anticipate this innovation in 1997 by arguing for the study of translation norms *as opposed* to simply accepting everything considered a translation as equivalent by default: "The interesting thing about the norms concept, and the issue of value raised by it, is that it invites us not only to recognize the primacy of difference but also to seek to explain the tenacity of equivalence" (65).

Why do we insist so emphatically on equivalence as the defining characteristic of all translation? Because, presumably, that's what we *need* out of translation: that's why we impose the regulatory translator-function on the translator, to restrict the proliferation of wild hermeneuses.

Certainly, that explains the utility of the GTP's presumed-equivalence principle in assessing, say, the translation of incoherently written source texts, as in the Second Strange Loop of Translation (pp. 13–16). The important thing in such assessments is not *enforcing* the equivalence requirement, but rather *using* that requirement to calm fears of translators' rampant hermeneutical excess. Global capitalism doesn't want to know how many liberties translators are taking—and, more to the point, doesn't want *societies* to know how much hermeneutical power translators wield—so long as that power is wielded surreptitiously and in the service of the translator-function, which is to say, regulating and restricting "the semantic potential and plurality of texts." As long as the translation of an incoherent source text *looks* proper, *looks* docile, and so can be presented and received as normatively equivalent, all is well.

The THP add a disclaimer:

> I do not think it necessarily follows from these remarks that translators should opt for different ways of translating. To my mind, the critical task of translation theory does not consist in advocating this or that "resistant"—or, for that matter, compliant or "fluent" or whatever—mode of translation. It consists, rather, in theorizing the historical contingency of these modes together with the concepts and discourses which legitimize them. The primary task of the study of translation is not to seek to interfere directly with the practice of translation by laying down rules or norms, but to try to account for what happens on the ground, including the ways in which translation has been conceptualized.
>
> (66)

And, we would add, the THP demons believe that the study of translation should try to account for the ways in which translation has been *regulated*, controlled, through the translator-function—and the ways in which it has escaped regulation, through the proliferation of difference.

The two other major points the THP make veer quite close to the DHP's notion of strange loops, and we think would have benefited from a reading of Hofstadter (1979/1989). The first comes in a section titled "Translation as an Index of Self-Reference":

> The specific way in which a community construes translation therefore determines the way in which individual translations refer to their prototexts, the kind of image of the original which translations project. The "anterior text" to which a translation refers is never simply the source text, even though that is the claim which translations commonly make. It is a particular image of it, as André Lefevere (1992) argued. And because the image is always slanted, coloured, preformed, never innocent, we can say that translation constructs or produces or, in Tejaswini Niranjana's words (1992: 81), "invents" its original.
>
> (60)

The "self-reference" of which translation thus arguably serves as an index is for the THP any culture's self-reference: how it not only understands its own identity but shapes that identity.

This observation, read dialogically with the DHP on strange loops, would expand the DHP's cognitive theory of rhetorical identity-formation from each human individual to whole human cultures:

> If this is true, then the whole cognitive and normative apparatus which governs the selection, production and reception of translations, together with the way in which translation generally is circumscribed and regulated as a certain historical moment, presents us with a privileged index of cultural self-reference. In reflecting about itself, a cultural community defines its identity in terms of self and other, establishing the differential boundary in the process. Translation offers a window on cultural self-reference in that it involves not simply the importation of selected cultural goods from the outside world, or indeed their imposition on others, but at the same time, in the same breath as it were, their transformation on the basis of and into terms which are always loaded, never innocent.
>
> (60)

Those "always loaded, never innocent" terms would, we suggest, be the THP's adumbration of the DHP's strange loops:

> Translation is of interest precisely because it offers first-hand evidence of the prejudice of perception and of the pervasiveness of local concerns. If translations were neutral, transparent, unproblematical, they would be dull and uninformative, either in themselves or as documents of cultural history and the history of ideas. They would be about as interesting as xerox machines. [Cf. the DHP's railing against literal translation as if it *were* a machine-like transfer of words without thoughts.] But because they are opaque, complicitous and compromised, the history of translation supplies us with a highly charged, revealing series of cultural constructions of otherness, and therefore of self. Being non-transparent, translations perhaps tell us more about those who translate than about the source text underlying the translation. It is the bias built into the practice of translation, the uses made of translation and the ways in which translation is conceptualized, that gives insight into how cultures perceive and place themselves.
>
> (60)

That series of adjectives, "opaque, complicitous and compromised," seems to us a nice approximation of strange loops. In Robinson (1997a: 115–16) we complained about the THP's failure to define the "opacity" of translation—a sight-based metaphor that didn't open to our understanding then—but it seems to us now that it is quite *clear* (another sight-based metaphor) what it means: one expects to be able to see through translation to the source text, or even to the intention of the source author, but one isn't. That inability to "see" through translation up to higher levels of the expected hierarchy of creation is of course one of the DHP's definitive characteristics of strange loops. (We'd had Hofstadter 1979/1989 on our shelf for several years when we wrote that response in 1997, but hadn't yet bothered to read it.)

The pairing "complicitous and compromised" in turn implies a moral or ethical recursivity in which every attempt to dissociate translation from bias and other forms of purposiveness only reveals its bias more egregiously. Sense-for-sense translation is supposedly the value-neutral (unbiased, unpurposed) reproduction in the target language of the transcendental (unbiased, unpurposed) meaning of the source text? That very pretense exposes not only the spin applied in order to conceal bias and purpose but the normativity of that spin (see Robinson 2011). For the THP it is precisely because norms are steeped in values that a norms approach to the study of translation is so revealing and so transformative. These would be the culturally normative strange loops by which otherness-becoming-selfhood is constructed: the strange loops from the DHP, the cultural norms and values from the THP.

The second point where the THP veer close to strange loops comes later in the paper, where they raise a cognitive complexity that becomes endemic to translation through level-shifts: "If we want to understand de Buck's translation of Boethius and communicate about it, we not only need to have a sense of de Buck's concepts and of his practice of translation, we also need to be able to *translate* his concepts and practice of translation into our translational concepts" (66). This is effectively Roman Jakobson's notion of intralingual translation as the channel of all communication; as the THP go on: "To understand and speak about someone else's translation, we must translate that translation. When we want to understand someone's discourse about translation, we have to translate that discourse and the concept of translation to which it refers. Our accounts of translation constitute themselves a form of translation" (66–7). We again assume that a translation will take us up the expected hierarchical levels, first to the translator, then to the source text, then to the source author. That will give us access to what the RBP demons call the past of the Author's book, or in this case the past of the translator's book, which we agree to pretend to attribute to the Author alone. Instead, we get more translations—or rather, we are ourselves pressured as readers into providing those intralingual translations. The THP comment:

> If descriptions of translation are performing the operations they are simultaneously trying to describe, the distinction between object-level and meta-level is rendered problematical. Descriptive translation studies in particular have been keen to keep object-level and meta-level well apart, but it turns out the object constantly contaminates its description. Even the scholarly study of translation is implicated in the self-description of translation as a cultural construct. There is a worm in the bud of descriptive translation studies and their claim to disciplinary rigor.
>
> (67)

This is Strange Loops 101. In the same year as *Le Ton Beau de Marot (Sans Boucles Étranges)* (1997), two years before the DHP's 1999 complaint that readers of *Gödel, Escher, Bach* had missed the point twenty years before, and a decade before *I Am a Strange Loop*, without giving the tiniest indication that they have even heard of the DHP or strange loops, the THP get it. "The object," they note, "constantly contaminates its description." The negativity of "contaminates" is also very close to the DHP: both thinkers are underscoring the *failure* of purification regimes, the DHP the regime of ontologizing psychology, which wants to reserve the meta-level for accurate empirical description of the psyche-as-object, the THP the regime of DTS, which wants similarly to create a scenario in which accounts of cultural systems are stable and empirical and rigorous and therefore accurate.

A more positive account of the *phenomenology* of strangely loopy translation, of course, still awaits its formulator. In retrospect, however, it is uncanny how convergently the DHP and the THP anticipated its possibility by undermining the methodological fictions that would seem to deny its necessity.

3

The Strange Loops of Translation as (Peri)Performative Identities

Let us begin this chapter by framing the famous liar paradox as a strange loop: if the Lying Cretan Pandemonium (LCP) are lying about lying, they're telling the truth; if they're telling the truth about lying, they're lying. Every foray we make into the loop in quest of *stable* truth, even if it's the stable truth of being an inveterate liar, brings us back to the unsettling destabilization of the beginning.

But now introduce Austinian performativity into the equation. Take that liar-paradox strange loop as a *performative dyad*: if the LCP are the "I" addressing us as "you," then our only recourse in the vortex of the loop is logic, which reaps the whirlwind. In Austin's terms, if the illocutionary force of the LCP's "I am always lying to you" is to tell the truth about lying and/or to lie about lying, the inescapable perlocutionary effect of that speech act is confusion. The best we can do performatively in response to the LCP's speech act is to name it a paradox: to quarantine it off from the set of logically "sensible" or "reasonable" utterances as an anomaly that cannot be parsed. Stymied, thwarted, trapped in the treacherous coils of formal logic, all we can do is charge the word *paradox* with a protective magic. So named, the liar paradox lies outside of rational discourse, and so need not be discursively engaged. It is now safely encased in glass in the Museum of Illogical Oddities.

But now take yet a third step: introduce into the LCP's speech act about lying and being a liar the periperformativity brilliantly introduced by the Eve Sedgwick Pandemonium or ESP (Sedgwick 2003: 67–91), what they call the "neighborhood around" the performative, the witnesses that ratify or fail/refuse to ratify the performative. If performativity is a dyad, the ESP argue—what "I" do to "you" with words—periperformativity is a multiplicity: how *they* ratify what "I" do to "you" with words.

So, for example, imagine *they*-witnesses to the LCP's famous *I*-declaration giving us-as-"you" advice:

- "Look at that grin, the sneaky bastard: he's having you on. It's not lying, it's just clever persiflage."
- "Yes, sure he means it, listen to him sigh, look at the downcast eyes, the self-deprecation is real. But that means the declaration rests on a truth-telling ideal. He's *deploring* his own lying, and the inveterate lying of his compatriots. You can trust him to be telling the truth about lying. There's a part of him that knows and admits it's wrong."
- "I know this Cretan. We go way back. He likes to say 'every Cretan's a liar' in order to carve out an exception for himself. The subtext is 'I'm the only honest one'!"

Now we are no longer limited to the sole recourse of formal logic. Now we are guided by social opinion. Now the *group* is entrusted with interpreting truth and falsehood. This guidance is not infallible, of course. This is the real world, where nothing is infallible. But at least now we're not trapped in the imaginary flatland of formal logic.

3.1 Logical Aporias and the Strange Loops of Periperformative Workarounds: Mauricio Mendonça Cardozo

Or take another case. Imagine the logical aporia as a strange loop:

> On ne parle jamais qu'une seule langue
> On ne parle jamais une seule langue. (Derrida 1996: 21)

> We only ever speak one language
> We never speak only one language. (Mensah 1996: 7)

An aporia built out of two structurally similar but mutually negating propositions isn't a liar paradox, of course, but it is a similar kind of strangely loopy rhetorical move, designed to elicit in readers a similar kind of feeling of being stopped, thwarted, unable to go on.

Again, however, that stop-sense is a logical performativity. What happens when we invite the periperformative witnesses to comment?

- "Look at that supercilious gleam in Derrida's eye. They think they're so clever! It's an intellectual parlor trick, designed to make them look smart and us look dumb."
- "What's this one language that gets always spoken and never spoken? An abstract kind of language, right? Communication raised to a high enough level of abstraction that it can take any form our imaginations assign it. Up there at that level there's no contradiction between 'just one' and 'never just one.'"
- "Who's this 'on'? Anybody? Nobody? Surely not everybody!"
- "Who are 'we'? You and I? All of us in this room? All humans? All living creatures? Everything in existence? Are rocks and rivers part of this 'we'?"

Implicitly, we suggest, those last two periperformative commentaries are pointing us to Mensah's English translation, and thereby to problems of translation. What the witnesses whisper to us is that the Jacques Derrida Pandemonium (JDP) don't say "nous ne parlons jamais." They use the impersonal "on," English "one": "One never speaks but one language" and "One never speaks only one language." And yes, French people often use "on" informally to mean "we," but is the rhetorical implicature the same? Are the JDP trying to include the same *type* of group, or trying to include them in the same way, with their "on" as the Patrick Mensah Pandemonium (PMP) are with their "we"? "We," after all, marks a self-recognized group—which may in fact involve group pressure to self-identify as a member. In Aristotelian terms, by saying "we" the rhetor(s) hope to construct themselves and their audience as a single coherent self-recognized group (saying "we" to that purpose of course involves what Aristotle calls ἦθος/*ēthos*). But then in the logical couplet above, who would be the aporetic rhetor? The JDP, the philosopher-demons who don't say "we," or the PMP, the translator-demons who do? We'll see in a moment that as translator-demons the PMP are supposed to lack the agency to "say" anything; but do they?

"One" is a different kettle of fish. "One" is an implicit and somewhat amorphous group, vaguely but determinedly organized by the surreptitious enforcement of ideological norms. "One" is an idealized representative of—and hintingly, indirectly, the speaker "oneself" is a self-appointed spokesperson for—the group of "right-thinking people." That would be the group of those people who *obey the rules*—the rules of that particular group, of course, which is implicitly being presented as universally exemplary. We've elsewhere (Robinson 2016a) characterized "one" as a first-person singular "they" or a third-person singular "we": "One doesn't do such things"/"On ne fait pas de telles choses" implies that some imaginary group of "them" doesn't do such things, but also that "I" belong to that group as well. "One" is not just a self-recognized group. "One" puts *normative but surreptitious* pressure on "one's" hearers or readers to belong to the group as well, or to act as if they belong to the group, by conforming at least outwardly to its norms. It's a passive-aggressive kind of rhetorical inclusivity, an act of persuasion whose implicit violence is difficult to identify and therefore even more difficult to resist.

Indeed, the ESP to the contrary, we would argue that "one," not "they," is the quintessential periperformative pronoun. In saying "one" one doesn't just witness, and doesn't just take an interpretive stand: one puts normative pressures on one's fellows to join one in the stand one has taken. "One" is an *organizing* pronoun. It works anonymously behind the scenes to bring coherent structure to a group.

So what are the JDP doing to us with "on"? As is well known, there are two different ways of understanding aporias: as (level 1) an actual impasse, and as (level 2) a rhetorical figure that performatively/hermeneutically *constructs* an impasse, which is to say engineers a *feeling* of impasse in the audience. On level 2 the aporia is typically understood in Aristotelian terms as serving (level 2a) λόγος/*logos* but is actually, we suggest, mobilizing (level 2b) an ἦθος/*ēthos*><πάθος/*pathos* convergence. Those would be levels, then, of what D.M. Spitzer (2019b: vi) calls "a betrayal of sense—sensation, the bodied alertness and responsion to the world—for the sake of reason": level 1, believing that the text possesses the stable (onto)logical quality of an impasse, without rhetorical situation; level 2a, believing that the rhetorical situation is designed (in the passive voice) to perform a logical impasse without the periperformative ἦθος/*ēthos* or πάθος/*pathos* of "witnesses"; level 2b, using the aporia to activate witnesses to produce the felt/phenomenological audience-effect of an impasse. Undoing the betrayal of sense, we believe, requires an exploration of the rhetorical situation in which the aporetic impasse is staged-and-experienced as ethical-becoming-pathetic or pathetic-becoming-ethical.

And how do we respond on level 2b? Not how *should one* respond: how *do we*? ("We" there is a rhetorical testing ground for the subliminal normativity of "one.") Do we cling cognitively on level 2a to the λόγος/*logos* of the JDP's rhetorical performance, resisting any subliminal affective pressures we may feel on level 2b to enter into and feel trapped inside a normative pathetic-becoming-ethical impasse? Do we distance ourselves pathetically from the intellectual force of the JDP's aporia-as-λόγος/*logos*? Do we secretly reject one side? ("I don't know about 'one,' but *I* never speak '*but* one language' … ") Do we try to find a middle ground that awkwardly encompasses both *On ne parle jamais*'s?

Or, to put these questions about rhetorical responses to Derridean aporeticity in (peri)performative terms, what we are asking is not "What are the illocutionary force and perlocutionary effect of a Derridean aporia?" but "How do we as the JDP's readers contribute to the viability of a Derridean aporia?"

This question is especially pertinent to a discussion of philosophy and translation—or to *Philosophy's Treason: Studies in Translation and Philosophy* (Spitzer 2019a), the title of the book to which we were asked to write a postscript (the very postscript that we are thoroughly rethinking and rewriting here)— because, as we began to hint a moment ago, the standard assumption about translation is that the translator has no textual agency and therefore is incapable of performing a speech act. Anthony Pym (1993), for example, makes this argument: citing the performative/non-translational speech act "2a. I declare the meeting open" as followed a half-second later by the "simultaneous" interpretation "2b. Je déclare ouverte la reunion," Pym notes that "although both utterances are well-formed performatives, only the chairperson's statement (2a) can properly perform. The interpreter's version (2b) will necessarily have a constative function with respect to the utterance that actually opened the meeting" (49). In fact, they say, the interpreter could justifiably render 2a as the "constative" utterance "2c. Le Président vient de déclarer ouverte la réunion" (49). And they comment:

> Thus, a properly translational relationship between two performative forms seems to imply that only one of those forms, the non-translational one, can actually perform. The second utterance, which arrives just a half-second too late, has its function blocked by the presence of an anterior first person, visible to all receivers of the translation. That is, the communication scene is already occupied by a first person with full capacity to perform. Chairpeople can open conferences; interpreters cannot. Or more generally, translational discourse seems by definition to exclude the possibility of a fully performative discursive function. And inversely, a translation that has a fully performative discursive function might then no longer be properly translational.
>
> (49)

The Anthony Pym Pandemonium's (APP's) binary is logically attractive, of course: "Chairpeople can open conferences; interpreters cannot." The problem with it is that its binary clarity depends on a pristine performative logic that is serenely undisturbed by the sociocultural phenomenology of periperformative witnessing: "The second utterance, which arrives just a half-second too late, has its function blocked by the presence of an anterior first person, visible [*by definition!*] to all receivers of the translation." In the post-Austinian corrective formulated by ESP (Sedgwick 2003) ten years after Pym (1993), those receivers have become periperformative witnesses—living human beings with the social power to ratify either "2a. I declare the meeting open" or "2b. Je déclare ouverte la reunion" as the performative utterance that opens the meeting. Perhaps 2a is neither visible nor audible to some of those witnesses; perhaps that group includes French-speakers who either don't understand 2a or just prefer to be guided by 2b. Some who opted for the performative power of 2b might well insist on its temporal priority, claiming that it was 2a that followed a half-second later. In the ESP's socioaffective model, that "full capacity to perform" that the APP *logically* assign the speaker of 2a is *sociologically* ratified by the audience, part of which may opt for 2b. As the ESP would insist, the APP too, like "Austin himself, ... tends to treat the speaker's agency as self-evident, as if he or she were all but coextensive—at least, continuous—with the power by which the individual speech act is initiated and authorized and may be enforced" (76).

Strikingly, in fact, even in asserting their rigid binary, the APP seem to be aware that things are actually quite a bit more complicated than that. The idealizations in their model, after all, fail to live up to the not-quite-binary approximations that pepper it: "Or more generally, translational discourse seems by definition to exclude the possibility of a *fully* performative discursive function. And inversely, a translation that has a fully performative discursive function might then no longer be *properly* translational." *Full* performativity and *proper* translationality would appear to be entailed by the crisp binarity of the APP model—but in practice neither seems to be easily attained. The APP want to exclude all middles, but can't quite manage it. As the ESP put it, one might want to explore the ways in which the philosopher's translatorial periperformatives dramatize "(what Neil Hertz refers to as) *the pathos of uncertain agency*, rather than occluding it as the explicit performative almost must" (76; emphasis added).

Following the DHP, we might want to picture the APP as suspecting that the procedural coherence of their meeting scene is threatened by strange loops in which temporal and logical priority become difficult or impossible to assign, and responding to that threat preemptively by imposing a rigid binary entailing a temporal order and logical hierarchy that are by definition—ontologically—beyond question. In that purified world, rather than threatening the dyads of performativity with strange loops, "the interpreter's version (2b) will necessarily have a constative function with respect to the utterance that actually opened the meeting." Not first but second—hence not performative but constative.

Studies of the translator's narratoriality from Hermans (1996) to Schiavi (1996) to Baker (2000) tend to be trenchant accounts of the strangely loopy agency found in the work of professional translators who are denied access to "certain agency" in their job descriptions but, like the PMP, find indirect ways of hinting at their agency nonetheless. These translatorial workarounds strikingly resemble the workarounds developed by the disempowered female characters the ESP track in Victorian fiction:

> Mrs. Glasher's periperformative solution, on the other hand, requires and solicits no demystifying deconstruction of agency by anyone else. "These diamonds, which were once given with ardent love to Lydia Glasher, she passes on to you": note the passive voice that elides the origin of the diamonds; note the coverture of first person by third person (where the explicit performative requires almost the opposite: the condensation of third-person forces in a first-person utterance); note the inversion in the word order of subject and object, so that the diamonds themselves already seem to acquire an oscillating and uncanny agency; note the double displacement (the diamonds "were given" to Lydia Glasher, but she only "passes" them along to Gwendolen) that foregrounds the material and legal problematics of how a woman may be said either to own or to transmit property.
> (77)

Because these are expressions of *uncertain* agency, it is extraordinarily difficult to sort out not only what is being done with language to whom, and whether it is a kindness or a violence or some other socioaffective impingement, but who is the subject/agent of the doing. That "coverture of first person by third person" in the ESP's account refers to Lydia Glasher speaking of "herself" in the third person—"she passes"—but that concealment of first-person agency in a distanced third person also hints indeterminably at the impersonal "one." Among the things that remain uncertain about periperformative agency is the degree to which the subject of any given utterance is socialized, collectivized. "2pp. One declares the meeting open."

It seems to us, in fact, that the strangely loopy uncertain agency of periperformativity runs through all of the chapters of Spitzer (2019a) like a dust-devil. For example, the Brazilian translation scholar the Mauricio Mendonça Cardozo Pandemonium (MMCP 2019) challenges the static binaries set up by Pagano and Vasconcellos (2003) between "inside translation studies" and "outside translation studies"—an in-group/out-group distinction that would of course be quite attractive to the members of any in-group. What such binaries ignore, the Cardozo Pandemonium (MMCP) argue, "is precisely the knowledge production and circulation in what Fleck [1986: 101] has called the 'exoteric circle,' which is made up of both laymen and specialists of other areas" (38): "the research [not only] *in* the area" but "*from* the area and *to* the area" as well (38). Like the strange loops that threaten the APP's meeting scene, Fleck's "exoteric circle" loops disciplinary (de)formations around into relentless border-crossings that are dissipative for disciplinary coherence but also generative for knowledge-exchange, revealing insiders as outsiders and outsiders as insiders and constituting the identity of "the field" as always-already (partially) other.

So, for example, the arm reaching from translation into philosophy or from philosophy into translation is neither "inside" nor "outside" translation studies, but rather a form of relationality, or of what the MMCP call "relational reason." And the (trans)disciplinary reach of all such arms tends to fragment what some might want to call the "inside" of translation studies: "In this sense, a specific philosopher or a poet may be absolutely central to one's research on philosophy or literary translation, while being at the same time of the most absolutely insignificance for a colleague working, for example, with an eye-tracking approach to translation process" (45).

And, we would add, while some of us are excited about a dozen or more such mediatory "arms" reaching between translation studies and other disciplines—for the DRP they would include hermeneutics, phenomenology, ordinary-language philosophy, socioaffective neuroscience, performance studies, postcolonial studies, Peircean semeiotic, cognitive science, religious studies, intellectual history, ancient Chinese philosophy, intercivilizational cofigurations, development studies, sustainability studies, transgender studies, and behavioral economics—that kind of shotgun interdisciplinarity/interrelationality is a *normative* red flag to some translation scholars reviewing submissions for publication. They not only insist that certain border-crossing books and articles shouldn't be published because they "aren't even translation studies," but exude the indignation of all "right-thinking people" at the authors' apostasy (betrayal of group norms): *one doesn't stray* from the strait and narrow of linguistic analyses of textual equivalence! To avoid the dissipative dangers of strange loops one must forgo the pleasures of transformative identity-(re)formation. Because the "insiderness" of this narrow disciplinary periperformativity does in some quarters still hold sway over the acceptable range of public performativity, it not only influences publishing decisions but also casts a pall over "outsiders'" view of what translation scholars do, as when Lydia H. Liu (1995: 1–42; 1999: 13–41; 2014: 149) dismisses translation studies *tout court* as fixated exclusively on interlingual equivalence (for discussion see Robinson 2017c: ix–xi and section 4 of this chapter, below).

The value of thinking through such matters periperformatively is highlighted for us in a passage from the MMCP like this:

> So, if we glance at home matters in the field of Translations Studies, we will notice the obvious: that there are such impressions of distance and proximity also within this very disciplinary circumscription. And in a case like the above mentioned, we cannot avoid admitting that the researchers of those two different branches of Translations Studies—one in the area of philosophy or literary translation, the other in the field of process-oriented studies—will very rarely read each other's papers, nor will they write to each other about the results of their projects. They simply don't have each other as their direct interlocutors. Then they don't even share the majority of their bibliographic repertoire, neither their technological or methodological resources. And the fact that they are both recognized as translation scholars—and assume translation as their research subject—doesn't prevent them to think translation clearly from different perspectives and in the pursuit of completely different interests. So, the least we can say here is that the inside-outside logic of the field of TS is hereby seriously undermined, at least from an epistemological, methodological, and technological point of view.
>
> (Cardozo 2019: 45)

Yes, that is "the least we can say here"—it is after all the main premise of the MMCP's chapter. But we would also want to underscore not just the periperformativity of these two specific groups of translation scholars not reading each other's papers, or sharing research results, or attending the same conferences, and so on, but the *conflicting periperformative normativizations* of these two approaches (philosophical/hermeneutic/literary and cognitive/empirical/process-oriented) *in and through and by these people doing what they do*.

3.2 The Strange Loops of Translating Heidegger's Untranslatables: Sabina Folnović Jaitner

The organizing question for this section and the next is (un)translatability: the strange loops of untranslatabilizing a text as translatable or of translatabilizing a text as untranslatable, where each -ize verb is specifically a periperformativity. The default philosophical strategy for untranslatability is, of course, to ontologize it—to betray sense for the sake of reason—so that, as in Cassin (2004) and Apter (2013, 2014), a given term or text either *is* or *isn't* translatable. The two chapters in Spitzer (2019a) that address themselves to Cassin's "untranslatables" both resist this tendency: Sabina Folnović Jaitner (2019, esp. 106–10) here in section 3.2 and Natalia S. Avtonomova (Gukasyan 2019: 14–19) in section 3.3. Arguing that when translated "philosophical untranslatables can also contribute to understanding of that very same, i.e., they can be understood as equivalent" (110), the Sabina Folnović Jaitner Pandemonium (SFJP) note that this perspective is made possible by none other than "the *reader* of philosophical texts" (110): the periperformative perspective. The SFJP then unfortunately neglect to explore *how* this reader reads, and *what* they are reading (just the words on the page, or other people's periperformative reactions to those words, or the group norms that have artificially restricted the hermeneutic options available to the reader?), what the status (normative? avant-garde?) would be of the kind of untranslatable-as-equivalent reading they favor, and so on. "It has previously been mentioned," they add, "that one of the characteristics of a philosophical text is the way in which it is read, be it original or a translation" (110). Yes, but "the way in which it is read" is an abstraction that addresses none of the periperformative dynamics that might *construct* "philosophical untranslatables" as "equivalent." Could we see an example?

Rather than wringing our hands while we wait, let's try constructing one ourselves. The SFJP cite Heidegger's *Dasein* as an untranslatable (108–9); let's see whether we can't unpack that into a full-bodied example of periperformativity, by slotting a simplified version of the MHP's argument in *Sein und Zeit* (1927/1967)/*Being and Time* (Macquarrie and Robinson 1962/2001) into what the ESP call the "I/you" dyad of the performative and the "one/we/they" of periperformativity.

First step: take *Dasein* as a kind of "I" that performs the key philosophical actions with which the MHP is concerned. Y'all may protest that *Dasein* is not a conscious subject in the way "I" is; but it is a practical engagement with the world, a human individuality that makes authentic choice and thus subjectivity possible. And once we notice that the "I" in the DHP's strange loops theory is not necessarily a conscious subject either—that it becomes conscious iteratively, through repeated analogically layered strange loops, and that its becoming-conscious is a process that may never reach full consciousness—we may note recursively (this is the burden of the third step) that Heidegger devotes considerable attention to the *submergence* of *Dasein*'s conscious "I" under the social dross of *das Man*, and to the desirable methods for smashing *das Man* and thereby bringing *Dasein* into full heroic consciousness.

Second step: take the "you" in the ESP's equation as "the world," especially other people. There is, of course, a problem with that, broached in the third step.

Third step: for the MHP the danger is that other people all too often gang up on *Dasein* so that it loses touch with its own inner nature, its true primordial ontology. What happens is that "other people" are internalized by *Dasein* as *das Man*, a transindividuality that is literally "the 'one'" but that Macquarrie and Robinson translate as "the 'they.'" This is problematic as a rendition of the MHP,[1] but useful in the context of this discussion, because as we've suggested the "one" is a more trenchantly periperformative pronoun for what the ESP call the "they."

So how would the SFJP suggest we read the MHP in English so as to construct the "untranslatables" *Dasein* and *das Man* as "equivalents"? If in the SFJP's tentative formulation "the way in which it is read" is a lectorial orientation to performativity, that lectorial subject-position might in the MHP's terms be either an "authentic" because "individualized" *Dasein* or an "inauthentic" because "collectivized" *das Man*. Which way should we go? Which pronoun should we mobilize for the "equivalentizing" understanding of Heideggerian "untranslatables"? We might take it for granted that the MHP would favor the "I," the reader's *Dasein*—but … would they really? We know that they rejected the literal English translation "being there." (This would have been well before Jerzy Kosinski's 1970 novel *Being There* and Hal Ashby's 1979 screen adaptation, in both of which the main character Chance is patently an embodiment of *das Man*.) What should be our takeaway from that ban on a literal translation? Was it the MHP's concern that literalism is too saturated with *das Man*, and thus with passivity? Should we assume, therefore, that the MHP would want us to translate (and read) as *Dasein*, as our own and y'all's own and each person's own authentic individuality? Or should we read that ban as implicating that for the MHP the optimum lectorial gravitation for the translator—or the reader of a translation, or perhaps even for any reader of the MHP even in German—is not authentically to their own primordial individuality but obediently to the MHP's wishes?

First binary gate: (1a) "authentic" self-directed lectorial/translatorial periperformativity vs. (1b) obedient MHP-directed lectorial/translatorial periperformativity. 1a would produce maverick translations and interpretations; 1b would reproduce the MHP's authorial will as slavishly as possible.

The dilemma across whose horns the MHP (or the Heideggerian acolyte) is stretched there is obviously that 1a is most in line with Heideggerian philosophy but might produce translations that deviate strikingly from the MHP's philosophy, while 1b, in seeking to produce accurate reproductions of the MHP's philosophy, seems to achieve that desirable goal by promoting a slavish submissiveness that is disturbingly reminiscent of *das Man*.

Second binary gate: (2a) surrendering boldly and heroically to the sovereign Ontological Interpretation as *channeled* by the MHP (i.e., understood as not invented or enforced by them) vs. (2b) surrendering slavishly to dogmatic Heideggerianism as to the "everyday interpretation" of an academic-philosophical *das Man*. 2a conveniently mystifies the MHP's authority over their own philosophical text as a kind of post-Romantic oraculism, so that it seems as if, in following the MHP, a self-styled hero were actually following their (we're tempted to say *his*) own primordial *Dasein*. 2b would be a becoming-demystified Heideggerianism colored by the increasingly widespread conviction, since Victor Farías (1991), and *a fortiori* since the release of the *Schwarze Hefte* (Black Notebooks: Heidegger 2014, 2019a, 2019b), that the brilliance of the MHP's philosophical thought is far too dogmatically grounded in Nazi ideology for comfort.

We think of the George Steiner Pandemonium's (GSP's) Heideggerian model of translation's "hermeneutic motion" in Chapter 5 of *After Babel* (1975/1998), where all translation is said to partake metaphorically of the violent raiding and plunder of foreign territories, the abduction, rape, and enslavement of foreign women—before some kind of mystical "restitution" (apparently based on Hölderlin's radical literalism) makes everything right again. The GSP's take on translation-as-violence is that, unpleasant as it may be, it is inevitable, because the MHP said that *all* interpretation is violent. Feminist translation scholars[2] have not been wildly happy with the GSP's explicit rape imagery, for obvious reasons, or even their apparent celebration of hermeneutic violence in general; but what if they're right?

It turns out that the MHP never claimed that *all* interpretation is violent. They only said that the *good* kind of interpretation, the *necessary* kind, is violent:

> Die *Seinsart* des Daseins *fordert* daher von einer ontologischen Interpretation, die sich die Ursprünglichkeit der phänomenalen Aufweisung zum Ziel gesetzt hat, *daß sie sich das Sein dieses Seienden gegen seine eigene Verdeckungstendenz*

erobert. Die existenziale Analyse hat daher für die Ansprüche bzw. die Genügsamkeit und beruhigte Selbstverständlichkeit der alltäglichen Auslegung ständig den Charakter einer *Gewaltsamkeit*. Dieser Charakter zeichnet zwar die Ontologie des Daseins besonders aus, er eignet aber jeder Interpretation, weil das in ihr sich ausbildende Verstehen die Struktur des Entwerfens hat.

(Heidegger 1927/1967: 311–12; emphasis the MHP's)

Dasein's *kind of Being* thus *demands* that any ontological Interpretation which sets itself the goal of exhibiting the phenomena in their primordiality, *should capture the Being of this entity, in spite of this entity's own tendency to cover things up.* Existential analysis, therefore, constantly has the character of *doing violence* [*Gewaltsamkeit*], whether to the claims of the everyday interpretation, or to its complacency and its tranquillized obviousness. While indeed this characteristic is specially distinctive of the ontology of Dasein, it belongs properly to any Interpretation, because the understanding which develops in Interpretation has the structure of a projection.

(Macquarrie and Robinson 1962/2001: 359)

The problem the MHP are addressing is that *das Man* teaches us to cover up our *Ursprünglichkeit* "primordiality," till we no longer feel it, can no longer channel it into action, indeed are no longer even aware that we have it. In order to uncover that mystical power, then, any "ontologische Interpretation"/"ontological Interpretation" must smash the cover, the false colors in which *das Man* has taught *Dasein* to wrap itself. "Die existenziale Analyse hat daher für die Ansprüche," the MHP write, "bzw. die Genügsamkeit und beruhigte Selbstverständlichkeit der alltäglichen Auslegung ständig den Charakter einer *Gewaltsamkeit*" (311)/"For the claims—or rather, for the complacency and the sedated self-evidentiality—of everyday interpretation, therefore, existential analysis invariably has the character of *violence*" (translation DRP). There is "die alltägliche Auslegung"/"the everyday interpretation" of *das Man* and there is "die ontologische Interpretation"/"the ontological Interpretation" of true primordial *Dasein*—and the latter must violently smash the former.

Third binary gate: (3a) a Heideggerian lectorial/translatorial periperformativity vs. (3b) a Sedgwickian lectorial/translatorial periperformativity. The GSP to the contrary, in 3a hermeneutic violence would be a necessary quality not of all translation but only of Nazi translation—translation designed to smash the Jews and other carriers of the deadly plague of *das Man*'s "complacency" and "sedated self-evidentiality." If, then, horrified at this vision of Nazi hermeneutics, we shrink from the MHP's violent metaphysical quasi-religiosity, we may decide that 3b as the collectivized witnessing articulated through the impersonal third-person pronoun "one" is the preferred alternative.

Then, rather than seeking to destroy *das Man* violently, we may come to value the power of the group mind to organize and normativize opinions into "truths" and "realities" while yet remaining open to challenge and resistance. Then, adopting *das Manian* periperformativity as our translatorial/lectorial orientation to philosophy, we may decide to render the MHP's "untranslatable" terms *Dasein* and *das Man* not with nouns but with noun phrases that more "commonsensically" explicitate the unspoken explanatory voice behind each German term: say, "What I Am" and "What One Does."

> Jeder ist der Andere und Keiner er selbst. Das Man, mit dem sich die Frage nach dem Wer des alltäglichen Daseins beantwortet, ist das Niemand, dem alles Dasein im Untereinandersein sich je schon ausgeliefert hat.
>
> (Heidegger 1927/1967: 128)

> Everybody's the other and nobody's themselves. What One Does, which answers the Who Does What question of everyday What I Am, is the Nobody Doing Anything to which all What I Am in its Being-among-one-another has already surrendered itself.
>
> (translation DR)

> Das Selbst des alltäglichen Daseins ist das Man-selbst, das wir von dem eigentlichen, das heißt eigens ergriffenen Selbst unterscheiden. Als Man-selbst ist das jeweilige Dasein in das Man zerstreut und muß sich erst finden—
>
> (129)

> The self of everyday What I Am is the What One Does-self, which we split off from the authentic, which is to say, authentically grasped self. As the What One Does-self is the temporally particular What I Am dissipated into What One Does, and must first find itself—
>
> (translation DR)

Notice what we're not doing there: playing with literalism. "Das Untereinandersein" is literally "the under-one-another-being," but we've translated it "idiomatically," using the idiom "commonly" considered "correct," "Being-among-one-another"—correct both as a Heideggerian technical term (as translated for philosophers) and as "standard" or "ordinary" English (as translated for nonphilosophers). "Unterscheiden" in the second passage is literally "to undersplit" or "to underpart," and we've translated that idiomatically also, as "split off" (though Macquarrie and Robinson 1962: 11 and Stambaugh 1996: 111 both have "distinguish"). We could multiply the implications of spatial metaphors to be possibly translated literally by considering an earlier passage as well:

> "Die Anderen" besagt nicht soviel wie: der ganze Rest der Übrigen außer mir, aus dem sich das Ich heraushebt, die Anderen sind vielmehr die, von denen man selbst sich zumeist nicht unterscheidet, unter denen man auch ist.
>
> (118)
>
> "The others" doesn't mean all the rest of them except me, from whom the I has expelled itself, so much as it does those from whom one doesn't split oneself off, among whom one also is.
>
> (translation DR)

The MHP don't use their neologism *das Man* here, but they're alluding to it, with "man"/"one" and "man selbst"/"oneself," obviously, and especially with "unter denen man auch ist"/"among whom one also is." Being "one" and "oneself" among "others" is precisely how *das Man*/What One Does is born.

Closer scrutiny reveals, however, that the MHP's spatial metaphors are *organized*—maybe ideologically. If to be "under" is to be mingled with the crowd, to be "over" ("die Übrigen" as the remainder, from "über," over) is to stand out from the crowd. Those not "undered" by or as What One Does are remaindered or "overed," as an overage. With that over-under imagery the MHP then mixes an inside-outside imagery: "*außer* mir, *aus* dem sich das Ich her*aus*hebt" gives us three outs, first "outside me" (myself as the inside, the others as outside), then the "I" heaving itself out of the others (the others as the inside, the authentic "I" as propelling itself to the outside). The overlapping spatial metaphors project visualizations of the relationships the MHP is trying to flesh out: not just *with* others or *alone*, but over-under others, inside-outside others. To be trapped inside others is to be under them, and so subordinate to them; to cast oneself off from them is to rise over them, to wield power over them.

In *not* translating those spatial metaphors literally, in other words, we are arguably impoverishing the MHP's message. But, one ("one"!) might want to say, so be it: translating spatial prepositions idiomatically, using common target-language usages, is What One Does. It is the professional marketplace norm. The MHP, by contrast, might want to say that true authentic *Dasein* would demand that we foreignize them, and in so doing smash the common idiomatic usages that "one" (the translator as the normatively "right-thinking" translator-function from Chapter 2) would prudently use. This then would be not only the MHP but

Berman and Venuti, too, as Schleiermacherian post-Romantic elitists who love power and despise the weak and the ordinary.

But note that the MHP's over-under and inner-outer metaphors *are* common usage in German, and therefore do not draw attention to the Nazi periperformativities the MHP may or may not be seeking to mobilize through them. Using those spatial metaphors idiomatically in German is again What One Does—even if in using them the What One Does-self is arguably weaponizing the primordial hermeneutic violence of What I Am to smash What One Does. The Heideggerian double-bind as strange loop: *Dasein*/What I Am circles around in quest of our authentic primordiality and the ideal striking position from which to bring the hammer down on *das Man*/What One Does, and finds that "I" have *become again* What One Does mouthing sniveling commonplaces and dead metaphors.

To foreignize those metaphors in English translation, especially to foreignize them radically, might seem to offer an escape from the strange loop; it would be to highlight them, and thus apparently to aggrandize *Dasein*/What I Am: "'The others' besays not so much the whole rest of the left-*overs outside* me, *out* of which the I self *out*-heaves; the others are much more those from whom oneself mostly does not *under*-part, *under* whom one also is." But, of course, maybe the MHP didn't even notice that they were using spatial metaphors in that patterned way; maybe that was us imposing a tendentious interpretation on entirely innocent metaphors in order to accentuate the Nazi power moves that those metaphors (may) sneakily adumbrate in German. And maybe, even if the MHP did use those over-under and inside-outside metaphors knowingly, and did want them to signal their Nazi politics to their ideological brothers-in-arms, they would not necessarily appreciate having them removed from their "natural" German context and aggressively denaturalized for all the world to see.

As the strange loops proliferate, who, in the end, is betraying whom?

3.3 The Strange Loops of "Good" and "Bad" (Periperformative) Translatabilities: Natalia S. Avtonomova and Tatevik Gukasyan

In their chapter of *Philosophy's Treason*, Natalia S. Avtonomova writes of the periperformative construction of translatability, and Tatevik Gukasyan translates:

Так или иначе, во всех этих процессах происходит выработка соизмеримостей, которые не даны нам от века: мы сами «строим соизмеримости», создаем промежуточные звенья, формируем пространство, опосредующее оригинал и перевод.

(Avtonomova unpub. 14)

One way or another, what is taking place in all these processes is elaboration of commensurabilities that are not given to us from the beginning: we ourselves "build commensurabilities," create intermediate links, form the space mediating the original and the translation.

(Gukasyan 2019: 27)

We do it. "We ourselves 'build commensurabilities.'" What is less clear is *how* we build them. People are involved—human actions, presumably social interactions, though the Natalia S. Avtonomova Pandemonium (NSAP) don't specify that. The "промежуточные звенья"/"intermediate links" that we create as a "пространство, опосредующее оригинал и перевод"/"space mediating the original and the translation" may or may not be twisted around into strange loops; our ruminations in this book would suggest that such loops would be almost unavoidable, but that may just be our paranoid reading (Sedgwick 2003: 123–46). Perhaps the NSAP's notion of how we "строим соизмеримости"/"build commensurabilities" is closer to the DHP's own, and thus far more straightforward and innocent than ours? What exactly is the "пространство, опосредующее оригинал и перевод"/"space mediating the original and the translation," and how exactly do we "формир[овать]"/"form" it? Is it a purely textual mediation, or are translators and editors and project managers also carving out problematic mediatory realms of "uncertain agency" in periperformative interaction with other human beings?[3]

The NSAP situate this discussion in the context of an intervention into the notion of "untranslatability," especially Cassin (2004) and the work the Emily Apter Pandemonium (EAP) have done translating (Apter et al. 2014) and commenting on (Apter 2013) that project. As translators themselves, the NSAP demons shy away from what the EAP (Apter 2013) nominalize as "the Untranslatable," a metaphysical positivity marked grammatically as a negative:

В общем мыслительном поле сама антиномия переводимость – непереводимость погружает нас в ситуацию неразрешимости. По сути, против непереводимости выступает и многовековая практика перевода.

Мне представляется целесообразным *отказаться от антиномии* перевода и непереводимости и считать перевод в принципе возможным, что, к тому же, подтверждается самим существованием межкультурных и межъязыковых контактов в течение многих столетий. Хороший пример подобного подхода дает нам Поль Рикер [Ricoeur 2004]. Он предлагает отказаться от спекулятивного противопоставления переводимости и непереводимости, которое он называет «метафизикой перевода». Исходить следует из того, что перевод существует фактически и затем уже рассматривать вопрос о том, каковы условия его возможности. Это аналогично установке Канта: он начинает с того, что естественные науки существуют, и лишь потом ставит вопрос об условиях их возможности.

(13–14)

In the common thinking field the antinomy of translatability–untranslatability puts us in the situation of unsolvability. In fact, the centuries-old practice of translation stands against untranslatability. I think it is more sensible to refuse the antinomy of translatability and untranslatability and consider translation basically possible, which is proved by the mere existence of intercultural and interlingual contacts for many ages. Paul Ricoeur [Brennan 2006] gives us a good example of the kind. He refuses the speculative opposition of translatability and untranslatability, which he calls "metaphysics of translation." One needs to start with the fact that translation exists in reality and then one can consider the question of conditions of its possibility. It is similar to Kant's assumption: he starts with the point that natural sciences exist and only after that he puts the question of their possibility.

(26–7)

Here phrases like "многовековая практика перевода"/"the centuries-old practice of translation" and "существование[] межкультурных и межъязыковых контактов"/"the mere existence of intercultural and interlingual contacts" hint rather abstractly at *what people do*: translators do translate. Faced with a difficult-to-translate word or phrase or idea, translators hired to translate do translate it. They do not say "Sorry, this is untranslatable." They come up with an approximation—a paraphrase, a footnote, a calque—that is only a "non-translation" confirming the mysterious presence of "the Untranslatable" to the radical purist for whom only perfect equivalence counts as translation. To that purist, of course, translation itself is impossible. There are no perfect equivalents, ever. Every translation is an approximation, a workaround, and therefore not really a translation at all.

One strange-loops take on that would obviously be deconstructive: the ways in which purity is constitutive of impurity and vice versa, and thus also untranslatability is constitutive of translatability and vice versa. We'll turn in section 3.4 to the theorists of Critical Translation Studies (see Robinson 2017c), who track periperformative histories of practical translation in order to show how those approximations and workarounds generate increasingly differentiated languages and cultures as source and target, and eventually generate purists who assert dogmatically that the two languages and cultures are incommensurable, untranslatable. But there is also the Hothead Translator Pandemonium for whom any claim of untranslatability is a red flag provoking them to translate that untranslatable text or phrase or word not just acceptably but near-perfectly.

In this section, however, let us take another strange-loops tack to the issue of (un)translatability by asking: how are some workarounds not only accepted but praised and admired as "great" translations, while others are condemned as "bad" translations? That crowd-sourced ability not only to distinguish between "better" and "worse" workarounds, but also to contest other people's assessments, is evidence that what is at work here is not a purely text-based up-down binary (translatable or untranslatable, equivalent or not equivalent) but a periperformativity. But what periperformativity? The NSAP seem to be struggling toward an engagement with something like that periperformativity, contra the metaphysical autisms of "the Untranslatable," but they don't have the vocabulary, or the theoretical framework, to cut to the chase. "Практика перевода"/"the practice of translation" and "межкультурные и межъязыковые контакты"/"intercultural and interlingual contacts" are a start, but remain at the level of vague adumbrations.

Another noun they invoke to hint at this socialized alternative, drawing on Jean-René Ladmiral (1986, 1994, 2000), is "опыт"/"experience," which they then clarify as "конкретные наблюдения"/"specific observations":

> Таким образом, мы все яснее видим, что априорные рассуждения о переводе ведут в никуда. В любом случае философский перевод – это такая особая абстракция, которая всегда требует определенной конкретизации, а потому любые рассуждения о переводе, как бы мы их ни называли (переводоведение, традуктология или традуктософия) теряют свой смысл, если они не опираются на опыт, на конкретные наблюдения, – лучше собственные, но пусть хотя бы и чужие. Но из этого, конечно же, не следует, что переводческая практика – сама себе голова и в философской рефлексии не нуждается.

(13)

Thus, we see more and more clearly that a priori reasoning about the translation leads nowhere. In any case, a philosophical translation is such a special abstraction that always requires a certain concretization and therefore any reasoning about the translation, no matter how we call it (translation studies, traductology, or traductosophy), loses its sense if it does not rely on experience, on specific observations—better its own, but at least others'. But this, of course, does not mean that the translation practice is its own head and does not need philosophical reflection.

(25)

The idea here seems to be that *we all know*—vaguely, but still—what "опыт"/"experience" and "конкретные наблюдения"/"specific observations" entail, so they don't need to theorize those nouns (or are they "the Untheorizable"?). Does "опыт"/"experience" really come down to just "конкретные наблюдения"/"specific observations"? "I" "observe" a problem while translating, or someone else "observes" a problem in our translation—those are experiences, yes. But is that "*experience*"? Is there really no more to it than that? And what *is* an "observation"? Just seeing? In Russian the NSAP's verb наблюдать does mean mostly visual engagement: seeing, watching, eyeing, and so on, but also overseeing, supervising, keeping watch over—like the marked-up translation file we just got back from an editor this morning. It can also be used more abstractly to mean realizing; but given that the NSAP specify "*конкретные* наблюдения," *concrete* observations, perhaps "realizing" would not be concrete enough. In English, "observations" can also mean verbal commentary on the seeing, but that semantic extension is not entailed in the Russian. "Конкретные наблюдения"/"specific observations" as an unpacking of "опыт"/"experience" would need to be restricted to visuals. The translator or editor or project manager or target reader (or someone else) *sees* the text, period.

So do the NSAP assume that a series of vague references to "переводческая практика"/"translation practice" or "опыт"/"experience" or "конкретные наблюдения"/"specific observations," as they themselves put it, "сама себе голова и в философской рефлексии не нуждается," or, as the Tatevik Gukasyan Pandemonium (TGP) translate that, "is its own head and does not need philosophical reflection"?

Such questions are, in fact, exceedingly difficult to answer in the abstract. As a "practical," "experiential," "observational" inroad to the periperformative problematic the NSAP seem to be attempting to uncover and elucidate, therefore, let us create a hypothetical periperformativity out of the TGP's

English translation of the NSAP's unpublished Russian original for this chapter. Remember that in a "pure" sense, from a God's-eye-view (which, needless to say, the DRP demons reject), all translations are by definition "bad" because perfect equivalence is impossible, and therefore the "success" or "failure" of any given translation is not an objective fact but an audience-effect, a periperformativity. So now, drawing on that, let us imagine that we wanted to convince y'all that the TGP's translation is "bad" or "good." What would we do?

The standard approach to this sort of task would be error analysis, of course. In order to argue (to begin with) that the TGP's translation is "bad," we would need to show that there are in it culpable deviations from the "correct" or "ideal" translation. This way of proceeding is common enough knowledge to be trivially true. What we want to suggest, however, is that underlying this highly idealized structural/textual/linguistic approach is an interesting kind of periperformativity. Let's see how that works.

In an error analysis of the TGP's translation aimed at proving it "bad," perhaps, we might want to begin by pointing out cases (all of these except the first taken from passages we've quoted above) where they guess at English articles and "miss":

1. "пройти «испытание иностранным»" (12):
    - "pass the 'trial of foreign'" (24; should be at least "pass the 'trial of *the* foreign,'" and indeed much better would be "undergo 'the experience of the foreign'")
2. "во всех этих процессах происходит выработка соизмеримостей, которые не даны нам от века: мы сами «строим соизмеримости»":
    - "what is taking place in all these processes is elaboration of commensurabilities that are not given to us from the beginning" (should be "*the* elaboration of commensurabilities")
3. "которое он называет «метафизикой перевода»":
    - "which he calls 'metaphysics of translation'" (should be "*the* metaphysics of translation")
4. "он начинает с того, что естественные науки существуют":
    - "he starts with the point that natural sciences exist" (should be "*the* natural sciences")
5. "переводческая практика – сама себе голова":
    - "the translation practice is its own head" (should be "translation practice")

It is relatively easy, though ultimately quite misleading, to say that there are *rules* governing the usage of English articles. For example, in 1 adjectives like "foreign" used as nouns always take "the." In 2 and 3 (and for that matter in 1) "the X of Y" structures always take the definite article. In 4, clearly bounded categories like "the humanities" and "the natural sciences" always take "the." And in 5, "translation practice" as translators doing things is an uncountable noun, so no "the"; with the definite article it might mean a specific translator's legal business entity ("I'm not asking about the royalties, I'm asking about the taxes on the translation practice").

The problem with explaining such "errors" with idealized structural rules, however, is that when we venture an explanation of what motivates those "errors," a focus on the rules takes us to cognitive deficiencies, which leads us to ask where the cognitive deficiencies come from, which leads us to deeper cognitive deficiencies, and so on *ad infinitum*. Any attempt to find a pragmatic bottom to that apparently bottomless series takes us to *culturally situated people*: since Russian has no articles, we may say (invoking "interference"), it's often hard for *Russian speakers* to get English articles "right"; and (therefore?) the TGP often stray from idiomatic English.

Note the implications there: we belong to one group, native speakers of English; and we are relegating the TGP to another, non-native speakers of English. (We assume, based on their name, that the TGP demons are Armenian, and fluent in the "colonial" "power" language in the Caucasus, Russian; but also that, because they're a university-trained intellectual, they're expected to be proficient enough in the "neocolonial" "power" language everywhere in the world, English, to publish in it. If Armenian is their L1, they are translating from their L2 to their L3.)

More than that, we seek to rope y'all too into the group of native speakers of English. Regardless of what language(s) y'all spoke in childhood, for our (hypothetical) argument to work we have to get y'all on our side. In Aristotelian terms, we appeal to ἦθος/*ēthos*: we are native speakers, and therefore possess unchallengeable authority; we speak, and y'all "naturally" (to the extent that y'all are either a native speaker of English or trust our authority) agree with us that a specific definite article or other target-textual rendering "should be" something different from what the TGP wrote. Our "ethical" address to y'all is designed to draw y'all into a periperformative What One Says or What One Does collective of native speakers of English, and ideally also of native speakers of English who can read Russian, and ultimately of RU>EN translators capable of assessing the

TGP's translation stereoscopically—who will validate our sense of "good" English and "good" RU>EN translation and condemn the TGP as a "bad" translator.

Next we would move to larger syntactic units:

6. "философский перевод—это такая особая абстракция, которая всегда требует определенной конкретизации":
   ➢ "a philosophical translation is such a special abstraction that always requires a certain concretization" (should be "is *the kind of* special abstraction that ...")
7. "как бы мы их ни называли":
   ➢ "no matter how we call it" (should be "*what* we call it")
8. "В общем мыслительном поле":
   ➢ "In the common thinking field" (should be "in the general conceptual field" or "in the general field of thought")
9. "сама себе голова":
   ➢ "its own head" (even literally this should be "a head all on its own" or "a head all to itself," but better would be, expansively, "a self-contained practical intelligence," in the sense of a φρόνησῐς/*phronēsis* [community of practice? Wenger 1998] that gets along perfectly well without abstract philosophical reflection)

And all through this periperformative witnessing we would try to pull y'all into the ἦθος/*ēthos*-becoming-πάθος/*pathos* of our normativized socioaffective orientation—first to the English language, then to the translingual passage from Russian to English—so that *we suffer together* through the TGP's translation. Our shared pain should stand proof of the intrinsic "badness" of the translation. The fact that we can analyze the sources of that pain linguistically helps, of course—helps us convince y'all that we're right, perhaps—but the primary evidence is somatic. The ESP, after all, theorize periperformativity in a book titled *Touching Feeling*: this is an *affective* theory of "witnessing" the performative utterance. It's not just that witnesses either ratify or prove unable to ratify the performance of a given speech act (in this case, the NSAP's article *in English*, the article *as if originally written* in English). It's that the πάθος/*pathos* of the "uncertain agency" of that witnessing is a socioecological blending of speaker/writer ἦθος/*ēthos* and listener/reader πάθος/*pathos*, *hidden transitively* in translator ἦθος/*ēthos*><πάθος/*pathos*, which we as a secondary (re)translator hijack consciously and verbally for our own ethical-becoming-pathetic purposes.

Also at work in that process, still in Aristotelian terms, is the strange loopiness of πίστις/*pistis*, which in Attic Greek is persuading, but also becoming-persuaded and so believing, and then persuading again: (1) the NSAP's authorial πάθος/*pathos* becoming (2) the TGP's translatorial πάθος/*pathos* as their English translator, then (3) the DRP's as the stereoscopic reader/witness and retranslator of the NSAP's original Russian and the TGP's English translation, and finally (4) y'all's as well, as our reader and periperformative witness. This would lead ideally, as Hans-Georg Gadamer (1960/1975; Weinsheimer and Marshall 2004 in English) would put it, to the fusion of all four πάθος/*pathos*-horizons in a single shared persuading-becoming-believing ἦθος/*ēthos* that we tacitly agree (or pretend) to attribute to the NSAP's unified Avtonomova-"I" alone as the official public performer of the speech acts (see section 1.3, above). We periperformative witnesses to those acts, translators and readers, don't count.

In this case, however, guided by our hypothetical intent to condemn the TGP's translation as "bad," the smooth (and mystified) flow of (1) through an invisibilized (2) to (4) is hijacked and derailed by (3), leading to the charge that their translation *thwarts* that potential sharing/fusion of πάθος/*pathos*-horizons and causes us as English readers/witnesses to refuse to ratify the NSAP's πίστις/*pistis* as ἦθος/*ēthos*—until as stereoscopic RU>EN witness we go to the Russian original and retranslate it, rethink it, reassess it, and finally ratify its performativity as (in Austin's term) "felicitous." The periperformative witnesses to the performance of a speech act—in this case first (3) we ourselves as the voice of condemnation, then (4) most of y'all as our willing followers—"ratify" or "fail to ratify" the NASP's performative through the socioaffective "ecology" (communal strange loop) of "believing" and/or "doubting."

And now imagine that we want to convince y'all of the opposite position, that the TGP's translation is truly excellent. How would we do that?

We would frame each of those nine ostensible "problem areas" as a deliberate and indeed stylized literalism, or let's say "foreignism." We would invoke the history of foreignizing translation theory—from the FSP (Schleiermacher 1813/2002) through the Antoine Berman Pandemonium (ABP 1984), which the NSAP cite in Russian, and the Stefan Heyvaert Pandemonium (SHP 1992, which the TGP do not cite in English, but never mind) to the Lawrence Venuti Pandemonium (LVP 1995)—in order to argue that by following the syntactic contours of the NSAP's Russian source text closely, the TGP infuse their translation with what Schleiermacher calls "das Gefühl des fremden"[4] (72), "the feeling of the foreign" (Robinson 1997/2014: 228). As we saw in section 1.4, that would be the feeling that we get when we read a text in a language that we don't feel entirely comfortable in, which the translator should *nachahmen* "simulate" in translating: the target reader, or at least some large group of demons in the Target Reader Pandemonium or TRP, reading in their first language, should get the feeling that they are actually reading in a foreign language with only middling competence.

As an example, look back to problem area 1, above: the NSAP's "пройти «испытание иностранным»," which the TGP rendered as "pass the 'trial of foreign.'" "Trial" is the first listing for "испытание" in a Russian-English dictionary, which may be why the TGP chose it; it is, however, arguably relevant to the TGP's choices that the NSAP footnoted that quoted passage *Испытание иностранным* as a Russian translation of the Berman Pandemonium's 1984 book title *L'Épreuve de l'étranger*, which in the SHP's 1992 English translation is *The Experience of the Foreign*. The SHP explain in their translator's note that the ABP took the title as a French translation of the Heidegger Pandemonium's (1985: 149) phrase "die Erfahrung des fremden"/"the experience of the foreign," though they also note that "The French *épreuve* has much richer connotations than the rather bland *experience*. There is a tinge of violence, of struggle, in it (captured best in the English *ordeal*), which makes it perfect as a rendering of Heidegger's *Erfahrung*" (vii). Actually, the ABP themselves (Heyvaert 1992: 158–61) say that the phrase "die Erfahrung des fremden"/"the experience of the foreign" comes from the Friedrich Hölderlin Pandemonium (FHP), whose translations the MHP was discussing in that piece.

We note also that in each phrase in the three non-Russian languages— German, French, and English—the syntax is "the X of the Y," not "the X of Y." In Russian, by contrast, not only are there no articles, but the instrumental case for the foreign(er)—иностранным—suggests that the syntax would be "X by Y," as in "trial by fire." The TGP do not essay "trial by foreigner," though that might have been an interesting *trial by a foreigner* in its own right! Another radically literal translation might have been "experience by foreigner," which, half-ignoring as it does the history of translation from the FHP to the MHP to the ABP, would likewise have been a striking foreignism precisely as a "trial by foreigner." Still, though the TGP's "passing the 'trial of foreign'" is neither idiomatic English nor radical literalism, it is of course quite *odd* in English, and therefore arguably foreignized. (The TGP are foreign to both Russia and the English-speaking world, and might well be celebrated as subjecting the English language to a "trial by foreigner" precisely by writing not that but "passing the 'trial of foreign.'")

This is, obviously, again a socioaffective/socioecological strange loop of periperformative "ratification." We pursue the TGP's translation strategies affectively around the loop, looking for enjoyable strangeness, ultimately looking for the brilliant translator behind the translation, drawing on this or that submerged translation history preiterating this or that choice in search of yet another way of admiring the oddness of their renditions—and every potential rising action leads us not to the image of a clever foreignizer carefully stylizing strange effects but to the flatness of the obvious, the interference error, the first entry in the dictionary. Because we are periperformatively determined to be witnesses for acclamation, for praise, we dive back into the text in search of ways to spin that flatness too in positive ways—and the process begins again.

In other words, it's not that "the text" has these specific linguistic or literary qualities, and that those qualities somehow mysteriously impact the target reader in predetermined ways—though unfortunately that is how the Berman and Venuti Pandemonia both tend to mystify foreignization. So, for that matter, does the Schleiermacher Pandemonium, quite often; but their explicit articulation of the necessity of simulating in the monolingual German reader/witness the periperformative "Gefühl des fremden"/"Feeling of the Foreign" does make it quite clear that foreignism is not a textual structure. It is a periperformativity, experienced not only affectively but *socioaffectively*, which is to say, through shared group feeling. The TGP's Russian is probably a lot stronger than their English, but that very imbalance helps them simulate for a target reader with no Russian at all the experience that an English-speaker with weak Russian might have reading the NSAP's Russian original. Just as Schleiermacher wanted German translators to translate foreign literature with the translational equivalent of a German accent, the TGP translate the NSAP with a Russian accent—and in that way excellently facilitate the target reader's "feeling of the foreign."

And as this reading of the TGP's translation suggests, it is a socioaffective periperformativity that is itself a site of contestation. Some readers "witness" it in one way, others in another. More: because periperformativity tends to conduce to normativization, it is also a site of conflicting normativities. It's not just that some people *like* foreignized translations and others *dislike* them: it's that the former group, following the AB and LV Pandemonia, seeks to normativize their approach as the only right way to translate and the only right way to read translations, and the latter group does the same, in the opposite direction.

That was the first takeaway from our two hypothetical takes on the TGP's translation: for the latter group any literalism is a "bad" translation; for the former group, any stylized translational awkwardness that leans toward literalism is a "good" translation. "Good" and "bad" translations are not stably binarized ontologies but competing periperformativities.

The second takeaway, of course, was that every attempt we make to hypostatize a periperformativity as an ontology mires us in strange loops. We keep prodding at the audience-effects that construct our "realities," turning corners in hopes of finding a pathway through to the ontology of what is—and just as we think we're succeeding, just as the massing clouds seem about to part and reveal solid objectivity, we find ourselves back at the start, mucking about in people's opinions, in the normativities of What One Does.

3.4 The Strange Loops by which Translation Shapes Collective Subjectivities: Sakai Naoki and Lydia H. Liu

In the three preceding sections, as the DHP predict, the audience-effects circulated rhetorically around strange loops are constitutive of professional/disciplinary identities: the disciplinary identity of Translation Studies in section 3.1, the professional identity of the translator as authentic primordial *Dasein* or scattered periperformative *das Man* in section 3.2, and the professional identity of the translator as "good" or "bad," competent or incompetent, in section 3.3. Here in this chapter's final section we are again concerned with periperformative identity-formation as a strange-loops audience-effect, but this time on a larger scale: the constitution of national languages and national cultures as bounded entities. As Lydia H. Liu (1999: 5) writes:

> We [may begin to rethink the idea of national languages and cultures] by trying to recapture a sense of the *radical historicity* of constructed—and often contingent—linguistic equivalences and nonequivalences that have emerged among the world's languages and societies in recent times. Such equivalences and nonequivalences in turn constitute the very identity of each national language and national culture. The emphasis on the *interactive and conflictual processes* rather than identities—again, not to be confused with the notion of hybridity or interculturality—may help fill some major gaps of knowledge in contemporary scholarship.

The fact that in this orientation to translation "the emphasis [is] on the *interactive and conflictual processes* rather than identities" does not mean identities are not important, but rather that it is out of "the interactive and conflictual processes" that "the very identity of each national language and national culture" is *constituted*. Translation Studies scholars assume the givenness of those identities; Critical Translation Studies scholars like Liu and Sakai track the emergence of national identities through regimes of translation.

Specifically, the Sakai Naoki Pandemonium (SNP) begin by distinguishing not so much between two *modes* of address, though that is how their model is sometimes described, but rather between two periperformativities that prestructure ways of thinking about modes of address—which is to say, that shape opposing performativities. The SNP's terms are "the *regime* of homolingual address" and "the *attitude* of heterolingual address"—"regime" and "attitude" signaling the organizing force of periperformative witnesses, "address" the I-you dyads of performativity. The regime of homolingual address is the normative periperformativity according to which each national culture and language forms a uniquely bounded entity within which all speakers instantly and easily understand each other, and *cannot* be easily understood by anyone from the outside. This regime, not only enforced but internalized and icotized as "reality," makes translation theoretically impossible. The attitude of heterolingual address, by contrast, periperforms-as-real the state of diffused difference in which we are all foreigners to each other and to ourselves, so that translation is ubiquitous. This is the SNP's counternormative periperformativity, according to which everything we say and everything we think is a translation; translation is repetition. In the periperformative regime of homolingual address, obviously, "the translator is supposed to assume the role of the arbitrator not only between the addresser and the addressee but also between the linguistic communities of the addresser and the addressee" (Sakai 1997: 14), and the fact that this arbitration is theoretically impossible effectively elevates the translator to the status of mythical hero. This, the SNP note wryly, tends to be the world-view that permeates Translation Studies. "Thus," they go on, "translation ceases to be a *repetition* and is rendered representable," so that "in the regime of homolingual address, translation as repetition is often exhaustibly replaced by the representation of translation" (14). It is through this representation of translation, too, that "collective subjectivity such as national and ethnic subjectivity" (14) is constituted. "Through the labor of the translator, the incommensurability as difference that calls for the service of the translator in the first place is negotiated and worked on. In other words, the work of translation is a practice by which the initial discontinuity between the addresser and the addressee is made continuous and recognizable" (14).

That "collective subjectivity such as national and ethnic subjectivity" would be the operative performative identity in this section: an identity not of authors or translators but of nations, cultures, language groups. What's interesting about the SNP's account of this outcome is partly that it is specifically periperformative-becoming-performative, the shaping labor of Sedgwickian "witnesses" coming to be seen, felt, and *realized* as a stable shape. Partly also, however, it is that the SNP frown on the outcome *as* that stable shape, which is to say *as a reality*: "In the regime of homolingual address, translation as repetition is often *exhaustibly replaced* by the representation of translation." Two discrete stages: translation as (heterolingual) repetition and the (homolingual) representation of translation. Once the former is "exhaustibly replaced" by the latter, it is no more. There is no more translation as repetition. In the regime of homolingual address, not only is the periperformative *attitude* of heterolingual address eradicated: the performative *reality* of heterolingual address is as well.

In other words, for the SNP, in the regime of homolingual address there are no strange loops of translation. As "the initial discontinuity between the addresser and the addressee is made continuous and recognizable" as a gap that is heroically crossed by the translator, our gaze rises from the gap to the mediator that spans the gap to the stable constitution of the cultural subjectivities on either side of the gap. There is no shunting of the gaze back to the initial incommensurability; there *is* no initial incommensurability. Rather than the "shift from one level of abstraction (or structure) to another, which feels like an upward movement in a hierarchy, and yet somehow the successive 'upward' shifts turn out to give rise to a closed circle" (Hofstadter 2007: 102), the upward shifts do indeed shift one upward. Rather than "wind[ing] up, to one's shock, exactly where one had started out," one climbs purposefully toward a stably objectified image of origins, causes, and contexts.

We would say, by contrast, that "in the regime of homolingual address, translation as repetition is *constantly generating* the representation of translation that is taken for a stable reality-shape." That is not only to say that even in the ideological normativization of homolingual performativity the periperformativity of audience-effects remains active, and indeed primary; it is also to suggest that the periperformativity of audience-effects is always coursing through strange loops. Let's see how that works.

The SNP's deceptively simple proposition is that "desire for 'Japanese thought' is invoked through the schema of cofiguration in the regime of translation" (1997: 51). The simple—but for Critical Translation Studies foundational—part of that is the definition of "the regime of translation" as "an ideology that makes translators imagine their relationship to what they do in translation as the symmetrical exchange between two languages" (51). In other words, the primal scene of Translation Studies, with individually unified speakers of two individually unified national languages who cannot understand each other and the non-pandemonial translator in the middle, heroically mediating between them, is an *imaginary.*

> The conventional notion of translation from English into Japanese, for instance, presumes that both English and Japanese are systematic wholes, and that translation is to establish a bridge for the exchange of equal values between the two wholes. A translation is believed to become more accurate as it approximates the rule of equal value exchange. In this regime of translation, it is required that one language be clearly and without ambiguity distinguishable from the other and that, in principle, two languages never overlap or mix like Siamese twins.
>
> (51)

In other words, the "primal scene of translation" as imagined by Translation Studies is not primal at all: it is created as a *"regime* of translation" through the ongoing strangely loopy iterations of "translation as repetition."

The deeper radicalism hidden in that proposition, however, is that "it is through this regime that, in the eighteenth century, the idea of [or 'desire for'] the original Japanese language was introduced into the multilingual social environment of the Japanese archipelago, where heterogeneous and creole languages were accepted" (51–2). In other words, it was through *translating* between "heterogeneous and creole languages" from the Japanese archipelago and other languages, such as Chinese, that the imaginary regime of translating between national languages began to organize first *thinking* about "the idea of the original Japanese language," then desiring it, and then, gradually, working periperformatively to purify that linguistic heterogeneity into, and ratify it as, a single standardized national language called "Japanese."

In what sense, then, are those regimes of translation strange loops? The Lydia H. Liu Pandemonium or LHLP (Liu 1999: 3) write of their desire to explore "both the familiar and not so familiar modes of value exchange in *global circulations*": it's not just that Japanese intellectuals translate from Chinese into some dialect or creole from the Japanese archipelago and linguistic unification somehow magically begins to happen, but that the values attached to specific ideas and images ("meanings") conveyed verbally circulate globally. To the extent that we expect to find strange loops in this theorization of translation, in other words, we should expect them to be global strange loops. "What I mean by the unfamiliar mode," the LHLP go on, "is the largely submerged and undertheorized forms of exchange such as the invention of 'equivalent' meanings between languages, struggles over the commensurability or reciprocity of meanings as values, and the production of global translatability among different languages and societies in recent times" (3). The LHLP demons are mainly interested in the translational production of such (largely unconsciously) negotiated engagements as equivalences, commensurabilities, and reciprocities, en route to macroengagements like "the production of global translatability"—but where are the strange loops?

In the LHLP model we detect (at least) two such loops, the "lock-in" loop that confirms expectations and so, through habitualization, idealization, and abstraction, produces tidy dictionary definitions as a phenomenological consolidation of clarified national languages and national cultures; and the "glitch" loop that disconfirms expectations and so, through surprise, deidealization, and diffraction, dislodges and resets the impulse to purify dictionary definitions into national languages and national cultures. This model aligns nicely with the DHP's parallels between an audio or video feedback loop and the constitution of the human "I":

> In this fashion, via the loop of symbols sparking actions and repercussions triggering symbols, the abstract structure serving us as our innermost essence evolves slowly but surely, and in so doing it locks itself ever more rigidly into our mind. Indeed, as the years pass, the "I" converges and stabilizes itself just as inevitably as the screech of an audio feedback loop inevitably zeroes in and stabilizes itself as the system's natural resonance frequency.
>
> (Hofstadter 2007: 186)

Translators working on translation jobs find patterns, collect patterns, build patterns into unconscious loops that come to seem like realities, *real* ontological commensurabilities between languages.

Leaving it at this, however, seems inadequate. The DHP, too, recognize that the difference between an AV loop and the human brain is that the AV loop doesn't *perceive*: "In any strange loop that gives rise to human selfhood, by contrast, the level-shifting acts of perception, abstraction, and categorization are central, indispensable elements. It is the upward leap from *raw stimuli* to *symbols* that imbues the loop with 'strangeness'" (187). In other words, it is the fact that translators *perceive* patterns in their work that makes the loops around which they circle in doing it strange. What we've called the "glitch" loop complicates things somewhat: translators also perceive deviations, paradigm-busting usages that force them to regroup and reframe the patterns they've been building—what the DRP (Robinson 1997/2020: ch. 11) call the "alarm bells" that force the habitualized translator to slow down and rethink things. But this all still sounds rather mechanical, like cognition operating on autopilot—or like what in section 3.3 the NSAP called "переводческая практика"/"translation practice" as "сама себе голова"/"a self-contained intelligence." (Part of the trouble is that the DHP have no idea what cognitive processes might cause a strange loop to "lock into" the experience of reality: yes, "By the time this brain had lived in this body for a couple of years or so, the 'I' notion was locked into it beyond any conceivable hope of reversal" (Hofstadter 2007: 188), but *how*? We essay an answer to that in Chapter 4.)

A more humanly complex loop is envisioned by the SNP (Sakai 1997: 16; bracketed numbers added):

> Hence, the figure of the Japanese language was given rise to cofiguratively, only when some Japanese intellectuals began to determine the predominant inscriptive styles of the times as [1] pertaining to the figure of the specifically Chinese, or as [2] being contaminated by the Chinese language. It is important to note that, [3] through the representation of translation, the two unities are represented as two equivalents resembling one another. Precisely because they are represented in equivalence and resemblance, however, [4] it is possible to determine them as conceptually different. The relationship of the two terms in equivalence and resemblance gives rise to a possibility of extracting an infinite number of distinctions between the two. Just as in the cofiguration of "the West and the Rest" in which [6] the West represents itself, thereby constituting itself cofiguratively by representing the exemplary figure of the Rest, [7] conceptual difference allows for the evaluative determination of the one term as superior over the other. This is how the desire for "Japanese language" was involved through the schema of cofiguration in the regime of translation.

First, there, Japanese intellectuals begin to distinguish some "styles of the times" *in the Japanese archipelago* as connected with Chinese culture *from outside the archipelago*, either (1) "pertaining to the figure of the specifically Chinese" or (2) "being contaminated by the Chinese language." It's not that those styles *were* Chinese; it's that they were increasingly perceived as either *pointing* to the Chinese or *contaminated* by the Chinese. They were, in other words, hybrid stylistic elements that had been thought of as vaguely local (not yet "Japanese"), but now increasingly came to be perceived as foreign. They were foreign influences that had been gradually incorporated into local styles—a cultural strange loop that had constituted a hybrid cultural reality.

Second, the intellectuals began to *translate* across the gap they had opened between the local and the foreign. The act of translating launched a new strange loop designed to consolidate vague perceptions of difference as stable cultural realities: the (3) equivalences posited by and for the act of translation also facilitated (4) the identification of difference, "giv[ing] rise to a possibility of extracting an infinite number of distinctions." This 3–4 stage is effectively the process theorized by the LHLP; what the SNP offer over and above the LHLP model is the tension between equivalence and difference in the organizing context of a perceived contamination. In the SNP's Japan the quest is not for what the LHLP, using the Moore and Aveling (1887) translation of Marx's *allgemeine Äquivalent* ("general equivalent"), identify as the capitalist quest for a "universal equivalent," but for a *nationalist differentiation* of Japanese styles from Chinese styles, which latter, falling out of the nationalizing view, leave behind a purified national identity. (To be clear, the Liu and Sakai Pandemonia are seeking out these unifications as desirable not *to themselves in the present* but to *others in specific historical contexts*. Their perspective is historicizing and descriptive.)

And third, what the SNP evocatively call the "schema of cofiguration" enables Japanese intellectuals not only to distinguish "Japan" from "China" and the rest of the world, but to set "Japan" up as (7) culturally *superior* to the rest of the world. Cofiguration is akin to Freudian negation and the Derridean logic of the supplement: by (6) "representing the exemplary figure of the Rest," the West (and by extension "Japan" as well) "represents itself, thereby constituting itself cofiguratively." Whatever is negated/supplemented as *not-Japan* drops away as the no-longer-pressing backdrop to the newly found wonderfulness of "Japan"; but the "forgotten" or denied first step (3–4) is the comparison in what the SNP call "the regime of translation."

What is strangely loopy about this "corrective" and "purgative" regime for the SNP is its cumulative and even anticipatory trajectory: the more one works translationally on "Japan and China" or "Japan and the rest of the world," the more differentiation one finds; and because what one is finding is *perception*, and because one is guided in the discovery of those perceptions by the expectations aroused by the cofigurative regime of translation, one keeps both anticipating finding more than was there before and finding more than one anticipates. And the more anticipatory/perceptual "evidence" one accumulates, the more "real" the differential identity of "Japan" (or whatever other cultural entity one is constituting) comes to seem—the more like a "natural" being that has always existed in precisely this form that one is "uncovering."

The SNP also reimagine not only the role but the subjectivity of the translator, first along the lines we explored in section 3.1, "based" (six years *avant la lettre*, in 1997) on the periperformativity broached in Sedgwick (2003):

> In the enunciation of translation, the subject of the enunciation and the subject of the enunciated are not expected to coincide with one another. The translator's desire must be at least displaced, if not entirely dissipated, in translational enunciation. Thus, the translator cannot be designated either as "I" or as "you" straightforwardly: she disrupts the attempt to appropriate the relation of the addresser and the addressee into the *personal* relation of first person vis-à-vis second person.
>
> (12–13)

We suggested above that the translator is periperformatively designated as "one"; the SNP don't go that far. In saying that "translation introduces a disjunctive instability into the putatively *personal* relations among the agents of speech, writing, listening, and reading" (13), however, they could be read as adumbrating the periperformative "one," as a first-person "they" or a third-person "we."[5]

Even more trenchant for a strange-loops reading of the cofigurative regime of translation, however, is the SNP's next suggestion:

> In respect to personal relationality as well as to the addresser/addressee structure, the translator must be internally split and multiple, and devoid of a stable positionality. At best, she can be a *subject in transit*, first because the translator cannot be an "individual" in the sense of *individuum* in order to perform translation, and second because she is a *singular* that marks an elusive point of discontinuity in the social, whereas translation is the practice of creating continuity at that singular point of discontinuity. Translation is an instance of *continuity in discontinuity* and a poietic social practice that institutes a relation at the site of incommensurability. This is why the aspect of discontinuity inherent in translation would be completely repressed if we were to determine translation to be a form of communication. And this is what I have referred to as the *oscillation or indeterminacy of personality in translation*.
>
> (13)

It is precisely because the translator "cannot be an 'individual' … in order to perform translation," and because this inability "marks an elusive point of discontinuity in the social," that the act of translation as a creation of "continuity at that singular point of discontinuity" is radically entwined in strange loops. It is "a poietic social practice that institutes a relation at the site of incommensurability," and that relation circles around *across* the "site of incommensurability," jumping the ditch, skirting the discontinuity, and in the DHP's image that means a reset, a return to inception. In the SNP model the "initial incommensurability [can be represented] as a gap, crevice, or border between fully constituted entities, spheres, or domains" only after translation, and thus "only retrospectively"; but, they add, "when represented as a gap, crevice or border, it is no longer incommensurate" (14). The *act* of translation creates "continuity at that singular point of discontinuity"; it is the "*representation* of translation" (14) that identifies it as a gap.

For the SNP, if there are strange loops of translation—not that the SNP demons ever mention such a possibility—they must be found in the attitude of heterolingual address. We find this unsatisfactory, however. Partly it's the fact that the performativity of homolingual address must constantly be *generated*, periperformatively, and that ongoing process must be strangely loopy. There is no stable reality to the regime of homolingual address. It is an audience-effect that only seems like an enforceable regime, or like the deep structure of reality, as long as the audience keeps suspending disbelief.

Partly also, however, as we argued in Robinson (2019: xii), both homolingual address and heterolingual address tend to invisibilize translation, the former by making it unthinkable, the other by making it as ordinary as breathing. We find unexplored promise in Sakai's notion of the translator as a "subject-in-transit," who, it seems to us, must be neither homolingual nor heterolingual but *translingual*— someone who addresses readers and listeners from within shifting transversalities in the fissures found between and within cultures. It is precisely the pandemonial translinguality of the translator that in the SNP model "marks an elusive point of discontinuity in the social," and that therefore in creating "continuity at that singular point of discontinuity" signally inserts translation into strange loops.

And if in the LHLP model there are potentially two types of strange loops, what we called the "lock-in" loop and the "glitch" loop, the translingual translator *is* the glitch. Like Vanellope von Schweetz (Sarah Silverman) in *Wreck-it Ralph* (Moore 2012), the translingual translator transports erratically, endangering the game—but in so doing also opens the possibility of transforming the game.

4

The Strange Loops of Translating Bodies

Translating bodies? What bodies? Surely we translate with our *minds*?

For the DHP, tellingly, strange loops too are disembodied. Their models for strange loops are typically technological (audio and video feedback loops), mathematical (Kurt Gödel's bizarrely powerful assault on Alfred North Whitehead and Bertrand Russell's *Principia Mathematica*), or abstract-hypothetical (the crashing on a frictionless pool table of imagined "small interacting magnetic marbles" [2007: 45] or "simms," which also "stick together to form clusters, which I hope you pardon me for calling 'simmballs'" [45]). Those "simmballs" are hypothetical abstractions standing in for symbols, which are, we suppose, putatively "real" abstractions whose strange looping-together gradually generates that "high abstraction behind the scenes" (179) called the "I": "In this fashion, via the loop of symbols sparking actions and repercussions triggering symbols, the abstract structure serving us as our innermost essence evolves slowly but surely, and in so doing it locks itself ever more rigidly into our mind" (186). The mental process enabling that "locking-in" is entirely mysterious for the DHP, but for convenience's sake they model it technologically: "Indeed, as the years pass, the 'I' converges and stabilizes itself just as inevitably as the screech of an audio feedback loop inevitably zeroes in and stabilizes itself as the system's natural resonance frequency" (186).

The hypothetical balls of "simms" or "simmballs" with which the DHP model the "dance of symbols" are "filled with meaning," they quip—but the whole point of modeling symbolic action with simmballs, we assume, is the need to protect the model against the kneejerk habits of readers looking for bodies, embodiment, situated embodied experience:

> the reason I say the simmballs are 'filled with meaning' is not, of course, that they are oozing some mystical kind of sticky semantic juice called 'meaning'

(even though certain meat-infatuated philosophers might go for that idea), but because their stately dance is deeply in synch with events in the world around them, [and] *will* continue tracking the world, *will* stay in phase with it, *will* remain aligned with it. (195)

No bodies, anywhere. Only "certain meat-infatuated philosophers" (and the rest of us poor uninformed sods who are naively happy with embodiment) are inclined to look for them anyway. The smart money is on abstraction.

One thing mammalian bodies do is feel, of course, and one might be tempted—we certainly are, and this chapter is organized around that temptation—to suggest that feeling might be the fuel driving the locking-in. But the DHP would demur: consciousness, including "what it feels like to be alive," is merely "the dance of symbols inside the cranium" (275). There is no affective force or fuel. "This dance of symbols in the brain is what consciousness is. (It is also what thinking is)" (276). It is also, presumably, what "feeling" is: just "the loop of symbols sparking actions and repercussions triggering symbols."

Other DHP demons, anticipating protests and complaints from skeptics, ask pointedly: "Who *feels* these symbols 'come alive'?" (276). "I suspect," some other "I am Doug Hofstadter" demon retorts, "that these skeptics would argue that the symbols' dance on its own is merely motion of material stuff unfelt by anyone, so that despite my claim, this dance cannot constitute consciousness" (276). We're not sure who those skeptics might be that are thus inclined to reduce the dance of symbols to "merely motion of material stuff unfelt by anyone": where would that motion be happening, and how could symbols possibly be "material stuff," and why would that stuff be "unfelt by anyone"? This claim is not something the DHP in the aggregate would claim as their own, of course, so presumably the "material stuff unfelt" argument is a straw-pandemonium; in any case their considered response is again to deny the embodiment of "feeling" by reducing it to awareness:

> The skeptics would like me to name or point to some special locus of subjective *awareness* that we all have of our thoughts and perceptions; I feel, though, that such a hope is confused, because it uses what I consider to be just another synonym for 'conscious'—namely, 'aware'—in posing the same question once more, but at a different level.
>
> (276)

If we *feel hungry*, we are aware of being hungry; if we *feel cold*, we are aware of being cold; if we *feel* "that such a hope is confused," we are aware of thinking that.

Feeling, therefore, is just more consciousness, which is just the purely mental dance of symbols (or "simmballs").

Again, in splitting their demons dialogically into two camps, two interlocutors labeled SL #641 and SL #642, the DHP has SL #642 insist on the importance of feelings but SL #641 argues that "those little sensual experiences are to the grand pattern of your mental life as the letters in a novel are to the novel's plot and characters—irrelevant, arbitrary tokens, rather than carriers of meaning" (285–6). Nice abstract reductionism. No body.

This concluding chapter is an all-fronts assault on such reductionist abstractions. Let's begin slowly, with an easy one.

4.1 The Strange Loops of Somatic Response: The DRP

All right, yes, the DRP demons are the collective author of this book. How then can we engage in dialogue with ourselves?

Different DRP demons answer that question differently.

One says that pandemonial individuals are always engaging in dialogue with "themselves"—some demons with other demons. Nothing new or unusual in that.

Another notes that the DRP wrote that book three decades ago, and most of the demons who wrote it are now dead. We read it with fresh eyes now—the fresh eyes of newer demons. If the cells in our bones are completely renewed every ten years or so, and our red blood cells every four months or so, how many demons can we have left from the writing of *The Translator's Turn* in the late 1980s?

Yet another (perhaps one smuggled in from the DHP—section 1.3) protests that we *share* demons, and that just as the mirror neurons simulate other people's (and pets', and maybe also plants') body states, so that their pain-demons and yawn-demons and happiness-demons and so on are reborn inside us, so too do reading-demons simulate the body states of the authors they read, so that author-demons ride into the reading pandemonium on the shoulders of reader-demons. This is presumably how the DHP found Carol-demons inside them after their death, and found DHP demons lurking in the eyes of a photo of Carol—and it's how the DRP internalized Finnish Source Author Pandemonium demons while translating the Mia Kankimäki Pandemonium's memoir, so that they partially formed a compound authorial/translatorial pandemonium (also section 1.3). It's how these dialogues have gone in previous chapters: the DRP internalizing author-demons while reading, and conducting each dialogue in part internally, sometimes speaking for their pandemonial interlocutors, sometimes speaking as them, sometimes setting up a conflict with them (initially felt inwardly). It's not as if the DRP were some kind of camera or recording device that recorded the written utterances of those authors mechanically, and simply played them back verbatim on the pages of this book! The dialogues are participatory on any number of levels, involving any number of pandemonial blendings. The DRP's engagement with the DHP operates mostly in this way as well, the DRP's demons admiring the DHP's demons so profusely that it's often difficult to tell from inside the DRP's body where one pandemonium ends and the other begins. When the two Doug-pandemonia were planning the one's visit to the other's classroom at Indiana University on February 25, 1997, one of them—we don't remember which—it hardly matters now—and at any rate it's a case in point—noticed the strangely loopy recursivity of the closing salutation "Thanks, Doug," with "Doug" naming either the sender or the recipient, or each in turn, in a duck-rabbit alternation, or both at once. "Both at once" would be the blurred or blended Doug-pandemonium, the sharing of Doug-demons. (When DRP demons find that DHP demons have fallen short—have failed or refused to theorize the body, for example—they don't feel indignant, or contemptuous, or triumphant: they feel ashamed, uneasy, anxious, as if they themselves had failed.)

Yet a fourth DRP demon recalls the photocopied precursor to *The Translator's Turn*, a bilingual book titled "Kääntämisen kääntöpiirit/The Tropics of Translation" (Robinson unpub.) that the DRP wrote in English and then translated into Finnish, experimentally, giving free rein to the Finnish-speaking and Finnish-writing demons that handled the translation, so that sometimes the Finnish text would deviate from its English "source" for eight or ten pages at a time, until other Finnish demons, or perhaps English-to-Finnish translator-demons, thinking stereoscopically, began to insist that the Finnish text be brought back into line with the English, and would keep up the pressure on the Finnish-writing demons until the two texts "merged" again, in a rough approximate sort of way that was always unstable, always susceptible to new "entropic" or "dissipative" deviations. When friends and colleagues expressed interest in reading that book, the DRP made fifteen or twenty photocopies,[1] with the Finnish on the verso and the English on the recto, and handed them out, and then, at one friend's urging, organized a two-hour colloquium on the book. When the book proved unpublishable—it was loosely based on *Éperons/Spurs* (Derrida 1978), but the 32-year-old DRP weren't the JDP, and Finnish wasn't French—the English demons rewrote it monolingually as *The Translator's Turn*, which proved considerably easier to publish. Still, elated as they were at the book being accepted by the Johns Hopkins University Press, many of the DRP's demons were sure the book would be ignored by translation scholars. It was too strange, too different, too far off the beaten path. But while top translation scholars did indeed attack it mercilessly, it caught on with younger scholars who were perhaps still more translators than scholars, and who felt alienated by the abstractions of high translation theory, and enthusiastically embraced the book's insistence on feeling and the body. Still, a young postgraduate student, being introduced to the DRP at a conference some time in the mid-1990s, exclaimed, "Oh, you're Doug Robinson! You're the guy who says translators don't have to think, they can just feel!"

And perhaps that exclamation will serve as a useful launching-pad for a strange-loops engagement with the book. The book's insistence on the conditioning and shaping power feeling exerts over thought is a problem in a culture that binarizes feeling and thought, obviously: though the DRP worked very hard in that book to address the dangers of that binary, and to warn readers against slipping into it, they did anyway, because the binary is hegemonic in Western thought and the Jamesian notion that feeling guides thought is not. William James, whose pronouncements on feeling and thinking are quoted and discussed in the book, has proved immensely influential for a whole school of cognitive and affective neuroscience, notably Antonio Damasio (1994, 1999, 2003), but the DRP didn't know that in 1987 or 1988, in writing the book—and in any case the mere fact that some influential thinkers and scholars and researchers have disputed the hegemonic binary for centuries carries no weight against hegemony. Common sense says feeling and thinking are opposites, so it must be true.

It is more or less along these lines that the DRP have typically whined about misreadings of the book. Until now.

For the DHP's strange-loops model now sets the stage for a very different kind of diagnosis. In this approach, the feeling-becoming-thinking or affect-becoming-cognition model advanced in *The Translator's Turn* might well be pegged as a hierarchical ladder leading linearly upwards from the despised lower levels of feeling/affect toward the respected higher levels of thinking/cognition.

If so, then commonsensical reader-demons, expecting the DRP to lead them up to that higher level and *stay* there, would be frustrated by our constant recursion to feeling. In the DRP's conception, feeling-becoming-thinking or affect-becoming-cognition is an Aristotelian entelechy that doesn't always reach the end it has within; but the commonsensical reading of the entelechy binarizes it too, so that potentiality is radically split off from actualization, which thus becomes the whole point of the process. The acorn is nothing: the oak is all. Feeling is nothing: thinking is all.

The DRP had been introduced to Charles Sanders Peirce by their dissertation director, Leroy Searle, and had published on the CSPP (Robinson 1985) before turning to the study of translation; but it would be a good ten or twelve years before they discovered the CSPP's emotional-energetic-logical interpretant triad, in which the entelechy climbing up from emotional interpretation through energetic interpretation to logical interpretation *starts over*—is forever starting over, in fact, in potentially endless strange loops brought only tentatively and pragmatically to a temporary halt by habit. That discovery formed the germ of this section's strange-loops explanation of the collapse of feeling-becoming-thinking entelechies into feeling-versus-thinking binaries: because the collective commonsensical commens-pandemonium believes that one climbs the ladder just once, and then rejoices in the finality of the promised end once one has reached the top, the fact that the loop keeps starting over is frustrating. Didn't we already dispose of all that irrational feeling nonsense and arrive at logical thought? Shouldn't we now *stay* there, and discuss translation as guided entirely by logical interpretants? If, in the terms of the DHP's definition of strange loops (2007: 101–2), the movement from feeling to thinking "feels like an upward movement in a hierarchy, and yet somehow the successive 'upward' shifts turn out to give rise to a closed circle," isn't that *wrong*? Isn't that to be *condemned*? If, "despite one's sense of departing ever further from one's origin," and seemingly moving in the right direction, toward logical thought, "one winds up, to one's shock, exactly where one had started out," back in the irrational swamp of feeling, surely the author is culpable, and must be held accountable for perpetrating that shock?

It was, we suggest, this frustrated hegemonic expectation of *culminating in thought* that provoked defensive dualistic readings of *The Translator's Turn*. That expectation made our insistence on keeping the whole feeling-becoming-thinking entelechy in view into a red flag: if for whatever perverse reason the DRP kept harping on feelings, and even seemed to be *subordinating* thought to feeling, that must mean that they *prefer* feeling to thought! That must mean they are actually saying that translators don't need to think, it's enough for them to feel! Translators can be stupid, uncritical, irrational—so long as their translational choices *feel* good! The whole book, in other words, must be preaching some touchy-feely 1960s gospel of doing what feels right, of throwing over all rational control of translation in favor of mindless gut instincts.

In this reading, in other words, the retreat from a Peircean emotional-becoming-energetic-becoming-logical-becoming-emotional-etc. loop to an emotion-versus-logic binary is driven by *frustration*—another affect. The condemnatory hegemonic attitude toward feeling that wants to dismiss it as pure random irrationality is justified as cognition but triggered by affect. More precisely, it follows the triadic entelechy from the affect of frustration through the conation of condemnation to the cognition of "understanding Robinson's book," but that entelechy is collapsed reductively into the *mask* of pure reason.

But now dive deeper. Where does the frustration come from? Why is that specific affect triggered? We suggested a few paragraphs ago that it is driven by what "the collective commonsensical commens-pandemonium believes": let's think about that for a moment. The "commens" is a Peircean coinage for a group mind (morphologically the Latin roots mean "with-mind") that makes communication possible:

> There is the *Intentional* Interpretant, which is a determination of the mind of the utterer; the *Effectual* Interpretant, which is a determination of the mind of the interpreter; and the *Communicational* Interpretant, or say the *Cominterpretant*, which is a determination of that mind into which the minds of utterer and interpreter have to be fused in order that any communication should take place. This mind may be called the *commens*. It consists of all that is, and must be, well understood between utterer and interpreter, at the outset, in order that the sign in question should fulfill its function.
>
> (Peirce 1992–8 2: 478 [1906])

CSPP scholars tend to understand the commens positively, as that self-organizing social ecology that facilitates communication among a group's members by bringing their minds into close normative conformity—what we call an icosis. But in addition to the icosis of communal harmony and understanding there is also an icosis of mob rule—not so positive. In this *Translator's Turn* case the icosis of common sense so effectively plausibilizes the feeling-versus-thinking binary as to block uptake of feeling-becoming-thinking entelechies—also not so positive, though not nearly as negative as mob rule. What is plausible to the cominterpretant or commens is real, is true, is valid, and what is implausible to it is a perverse distortion. And what happens if someone *harps* on such perversely distorted implausibilities? What happens if someone keeps trying to make one recognize the limitations of one's commonsensical participation in the plausibilizations of the commens, even after one has plainly stated one's utter rejection of their blandishments? Why, one gets frustrated, of course. One gets angry.

One resorts, in other words, to affect. Why? Because, we suggest, the commens is itself created and defended affectively. Because it is a socioaffective ecology that works on us affectively, and therefore unconsciously. Because we *lock into it*, as we do into any icotic "reality," through shared evaluative affect—precisely what the DHP abstraction-demons deride as the mystical fantasies of "certain meat-infatuated philosophers."

We wonder, though: surely it's beyond question that the DHP loved their wife Carol? Surely the sharing of demons between them, and thus also the composite being that their love created, and that survived their death, was saturated in affect? Surely the DHP would not want to disavow or demean the affective embodiment of their shared being by ridiculing it as "meat-infatuated"? Both "meat" and "infatuated" are, of course, satirical terms, terms of ridicule, "meat" serving to ridicule embodiment and "infatuated" to ridicule affect. It seems to us a given that the DHP would not want to describe their love for Carol as an "infatuation" with their "meat." But what, then? Remove the bitter angry satire directed at "certain philosophers" and say that *the DHP loved Carol's embodied being*: have y'all said anything substantially different? If not, what is served by selectively satirizing a love of embodiment in "certain philosophers" while cherishing that love of embodiment in one's marriage?

But we also wonder whether we aren't in fact being unfair to the DHP, who also write, elsewhere in *I Am a Strange Loop*:

> In the world of living things, the magic threshold of representational universality is crossed whenever a system's repertoire of symbols becomes extensible without any obvious limit. ... —and this is how stories work, including novels, movies, etc. Although I have been depicting it somewhat cynically, representational universality and the nearly insatiable hunger that it creates for vicarious experiences is but a stone's throw away from empathy, which I see as the most

admirable quality of humanity. To "be" someone else in a profound way is not merely to see the world intellectually as they see it and to feel rooted in the places and times that molded them as they grew up; it goes much further than that. It is to adopt their values, to take on their desires, to live their hopes, to feel their yearnings, to share their dreams, to shudder at their dreads, to participate in their life, to merge with their soul.

(2007: 246)

Clearly, there are DHP demons that believe in the primacy in human life not only of affects like values, desires, hopes, yearnings, dreams, and dreads, but of sharing those affective linkages with others, "being" those others. The question, therefore, is not so much whether the DRP demons are being unfair to the DHP but whether some DHP demons are being unfair to other DHP demons, in ridiculing "meat-infatuation" and dismissing affects as "irrelevant, arbitrary tokens, rather than carriers of meaning."

Yes, academic writers are supposed to be "consistent." Logical inconsistencies and outright contradictions are bad form—so bad that for some purists they invalidate the entire argument. But *our* argument throughout this book, along with the DHP's arguments for the strange-loops model on which it is based, would dismiss that kind of purism as the worst kind of utopian idealism. We are tempted to quote the Ralph Waldo Emerson Pandemonium to the effect that "a foolish consistency is the hobgoblin of small minds," and the Walt Whitman Pandemonium along the same lines: "Do I contradict myself?" they asked grandiosely in *Leaves of Grass* (1855/2009). "Very well, then ... I contradict myself. / I am large ... I contain multitudes" (63). But we don't need to return to the American Renaissance to find a model for this kind of multitudinous "largeness": the DDP's Pandemonium model of human consciousness offers a more recent neurophilosophical justification as well. And if y'all have been rolling y'all's eyes at our plural theys and y'alls and our DRP and DHP and DDP abbreviations, wishing we would just give it up and go back to referring to individuals in the singular, here is tangible evidence of the value of proceeding as we have been proceeding.

Unless y'all would rather dismiss the DHP's argument *tout court*? Or, perhaps, spin what we take to be inconsistencies among the DHP's demons as not inconsistent at all, but based on our egregious misreadings? Either way, be our guest.

Now let's take a roundabout looping path back around to such questions by engaging Henri Meschonnic in dialogue about rhythm.

4.2 The Strange Loops of Knowledge-Translation as Mouthable Rhythm: Henri Meschonnic[2]

In this section we expand the scope of translation along the lines of the SNP's "attitude of heterolingual address" (section 3.4)—translation as ubiquitous—by rethinking "knowledge-transfer" or "knowledge-exchange" as "knowledge-translation." Knowledge-transfer is, of course, a term for the sharing of knowledge either from one part of an organization to another or from one societal institution (such as a university) to another (such as an industry or an NGO); knowledge-exchange is a broader term covering basically the same thing, but specifying that the "transfer" is not unidirectional, and may even be circulatory.

What, then, would the shift to "knowledge-translation" imply? Given the deliberate push past purely verbal communication in both knowledge-transfer and knowledge-exchange—the conception of "knowledge" as both highly embodied and highly communalized, as in Wenger's (1998) notion of Communities of Practice, where knowledge may be practical knowhow internalized largely unconsciously through interactions with others—knowledge-translation would have to integrate embodied/communalized notions of *knowledge* with embodied/communalized notions of *translating*. But what would that entail?

We want to suggest that the most comprehensive model of fully embodied and communalized translating has been offered by the Henri Meschonnic Pandemonium (HMP). Let's build up to that offering thoughtfully.

In "The Meaning of Rhythm," suggesting that "rhythm may provide us with an opportunity newly to understand the relation between language and the body," the Amittai Aviram Pandemonium (AAP; Aviram 2002: 161) note that the traditional view of rhythm as "the regular occurrence in time or space of a foregrounded event" has recently come under fire from several pandemonial theorists, especially the pandemonia of Julia Kristeva, Philippe Lacoue-Labarthe, and Charles Bernstein, as well as the HMP. All of these thinkers, the AAP point out, rely heavily for their rethinking of rhythm on the Emile Benveniste Pandemonium's (EBP 2's) new look (in 1951) at the etymology of rhythm as derived by the Emile Boisacq Pandemonium (EBP 1, 1916) from *rheo* "to flow." Traditionally, according to the EBP 1, rhythm is derived from the flow not of a river, say, but of a sea, and specifically of the pounding of waves on a shore. Obviously, waves *flow* through water, and when that flow is interrupted by a slope of land rising up beneath it, waves crash rhythmically on the beach;

but as the EBP 2 (Meek 1971: 282) point out, the ancient Greeks never used the verb *rheo* of the sea. They also never used the noun *rhuthmos* of waves—or even, in the oldest texts, of what we now call "rhythm." What the Greeks took to flow was specifically a river—and obviously a river has no rhythm. The EBP 1's authoritative etymology, the EBP 2 insist, was based on "pure invention" (282).

How, then, was *rhuthmos* used? Leucippus and Democritus use *rhuthmos* or *rhusmos* to mean something like "form" or "configuration"; the EBP 2 (Meek 283) proffer a long and detailed list of further examples, in all of which *rhuthmos* is "understood as the distinctive form, the characteristic arrangement of the parts in a whole." But they also freely admit that the derivation of *rhuthmos* from *rheo* is quite accurate; the touching point between *rhuthmos* as form and *rhuthmos* as flow is that the ending *-(th)mos* is typically used not for static abstractions but for "the particular modality of its accomplishment as it is presented to the eyes" (285), so that *rhuthmos*

> designates the form in the instant that it is assumed by what is moving, mobile and fluid, the form of that which does not have organic consistency; it fits the pattern of a fluid element, of a letter arbitrarily shaped, of a robe which one arranges at one's will, of a particular state of character or mood. It is the form as improvised, momentary, changeable.
>
> (285–6)

And as the AAP (Aviram 2002: 162) explain:

> It is in the words of Socrates that this form or shape of a moving body is required to follow "measure" and order—in other words, to be metrical—so that in Plato the term occupies the exact point where the ancient and modern senses of the word overlap. After Plato, apparently, *rhythm* has meant increasingly what it means today. From the point of view of these post-Benveniste theorists, the conventional meaning of rhythm today is informed by *measure*, what Plato calls *metron*. These theorists wish to return us to the pre-Socratic sense, where, they believe, rhythm is closely bound up with subjectivity and discourse.

The AAP are not fans of this new radical approach, which is, they note at some length, mired in historical and philosophical ironies:

- The EBP 2 approach to rhythm, reaching back to a pre-Socratic meaning as "truer" or "more authentic" than the modern philosophical rigidifications imposed by the Plato Pandemonium (PP), is a Heideggerian (MHP) move, stretched in the history of ideas between Hölderlin's notion that "rhythm is the normality of the subject, and the caesura is the revelatory moment," and Nietzsche's insistence that rhythm/music is the Dionysiac, which "affords a break in the subjectivity of thought" (Aviram 2002: 163). And if rhythm therefore plays the role of "philosophy" in this MHP morality play, and the caesura plays the role of "the place of thinking the grounds of philosophy," the AAP note, "the break in thinking is still thinking. There is nothing outside thought. What would be the unthought of thinking about the unthought?" (163)

- What the PP mean by *rhuthmos* is not what we mean by rhythm, which for PP would have been complexly tied up with poetry=music=dance. "An attempt in modern times to connect the shape of a moving being with subjectivity," the AAP (164) say, "presupposes a conception of subjectivity which attends that very semantic shift over history."
- The PP's notion that a *regulated* beat in music will shape the citizenry in regulated ways is in fact the first modern bridge between rhythm and subjectivization. "Rhythm is not so much a sign of subjectivity, as it would be in Meschonnic, as subjectivity is a sign of rhythm" (Aviram 2002: 164). This also means that rhythm and subjectivity become circular in their effects, mutually constitutive, to the point where it should in theory be impossible to stand far enough outside of rhythmically induced subjectivity to judge or determine the proper regulation of musical beats for the education of citizens in the Republic. (Let's call this the HMP Strange Loop #1.)

What does all this have to do with knowledge-translation? To begin to answer that question let's take four key steps past the AAP's useful formulations: (A) translation is rhythmic, and translational rhythms are (B) intersubjective, (C) serial, and (D) experienced by the body-becoming-mind. Step A answers the question about the relevance of embodied rhythms to knowledge-translation; step B communalizes that embodiment; C helps refine A, and D helps refine B.

A. We can think of this initial postulate in at least two ways—say, a weak and a strong version. The weak version would be that we are only interested in translations that can be thought of as rhythmic, or rhythms that can be thought of as translational. A weakly rhythmic definition of translation might focus, for example, on translating verse as verse—say, the DHP's translation of *Eugene Onegin*. The strong version by contrast would insist that all translation is rhythmic, and possibly even that all rhythms are translational. We will find the HMP embracing the strong version, and insisting that translations that care nothing for rhythm are not so much arrhythmic but rather dysrhythmic, organized around ineffective rhythms. The strangely loopy (layered-analogical) equation would then be body=rhythm=translation=body, and around and around.

B. If in HMP Strange Loop #1 rhythm subjectivizes and subjectivity rhythmicizes, around and around, each does so *communicatively*, which is to say intersubjectively/inter-rhythmically. Rhythm intersubjectivizes, in the sense of creating/channeling/interrupting/transforming subjectivities both within and between speaker and hearer; subjectivity inter-rhythmicizes, in the sense of insinuating its kinesic expressions into the turbulence of two or more bodies cross-communicating. Rhythm and subjectivity are thus both *transfers*, or *translations*; and to the extent that the (inter)subjectivity that is (inter-)rhythmically transferred back and forth is a form or *rhuthmos* of knowledge, rhythm is a—some would argue *the*—primal channel of knowledge-transfer or knowledge-translation:

> Lacoue-Labarthe is right that rhythm both shapes and breaks thought. But rhythm is the caesura of thought, as in poetry, precisely because rhythm goes on. A caesura in poetry only strengthens our anticipation of the rhythm that will return both to give forth and to disrupt the play of meaning in imagery. Poetry, then, does what all knowledge, including scientific knowledge, does: it makes sense of a world by representing it, but contains within it the signs that the world is not representation and thus cannot allow a representation to be found that is perfectly commensurate with it. The failure of knowledge is also its success, because it is what enables us to be free.
>
> (Aviram 2002: 170)

C. Thinking knowledge-translation as rhythm temporalizes it, despatializes it—helps us remember that we live in time and so tend to process and internalize knowledge one rhythmic beat or caesura at a time. This would be what the HMP call "serial semantics," their act of resistance to the λόγος/*logos* as spatialized meaning, laid out like a map—or as spatialized rhythm, graspable at a glance:

$$- / - / - / - / - /$$

One quick glimpse of this visual pattern has us muttering "iambic pentameter." This spatialization of rhythm is in fact something like the conception that the AAP defend, somewhat problematically:

> But if architecture is rhythmic, it is also spatial and still. In this, is architecture not unlike music and dance, which are never still but always moving? But the

rhythm of music and dance also bring about a stillness as well. Although the body of the dancer moves in time, the beat remains the same. What does it mean to say that the beat remains "the same"? The anticipation of the beat, which stands in a stillness before us, matches the memory of the beat, which stands in a stillness around us, beneath us. It is only if, say, the tempo momentarily changes that the stillness is momentarily disrupted. Even if such a tempo forms part of a larger pattern that can be anticipated, then we return to stillness. Rhythm allows us to move with great energy but to remain still and serene all the while.

(168)

The HMP would deny this, indignantly. The relative predictability of a regular rhythm is *not* a stillness; nor is our participation in that predictability. It is precisely because a regular beat *can* change that we cannot participate in a rhythm with serenity. Every regular beat is a confirmation of our expectations, but the fact that the phenomenology of music—of participation in rhythm—requires the repeated confirmation of expectations grounds it in insecurity, inserenity, in the fear of broken expectations.

D. As the HMP also remind us, rhythm is as much a bodily experience as it is a cognitive one. "The break in thinking," the AAP say, "is still thinking"—but perhaps it's also *more* than thinking?

The HMP define good translation, true translation, in terms of the poem, and the poem in terms of rhythm, and rhythm as the body language of the subject (Meschonnic 1982). The interesting questions raised by that definition are: what is the subject, and what is this body it apparently has that possesses or produces language-as-rhythm?

The translatorial HMP (Meschonnic 1970, 1981, 2001, 2002, 2003, 2005) were primarily focused on the Hebrew Bible, the most authoritative text of which—the Masoretic—has the rhythmic accents marked in the text as performative indices, to guide its reciters in reading it aloud, much the way a modern actor will mark up their lines in a script for rhythmic, tonal, and other prosodic features. The Hebrew term for those accents that were added to the biblical text by the Masoretes is *te'amim*; and as the HMP (Boulanger 2011: 71; see also 99, 119, 136, 143, 163) never tire of reminding us, the singular form of that word, *ta'am*, means literally "the taste in one's mouth, flavour, flavour being the very reason for the act of saying, and it is first and foremost the meaning of orality. What comes from the mouth." "When this goes through translation," the HMP write elsewhere, "it has to be mouthable" (133). "To *taamicize*," they add, "is to oralize, in the sense where orality is no longer sound, it is subject, it is thus translating the power of a language, and no longer just what words say" (141). Hölderlinizing textual signals of rhythmic accent as taste or flavor or mouthable orality would seem to indicate a desire to find a way from textuality or discursivity to the source-reader bodies that savored the Hebrew Bible between two and three millennia ago, and thus to model a corporeal response to the "poem" for target readers today; but the HMP is not particularly forthcoming on the exact nature of the "body" whose rhythmic language they take to be our primary indication of the textualized subject. They mostly seem inclined to invoke what they call the "body-in-language continuum"; but we are suggesting that the HMP are actually thinking of body, subject, and language in terms not of linear continua but of strange loops. For them the poem is "an invasion of the body and its power in language[:] not flesh, but maximum rhythmicization" (132). "Maximum rhythmicization" recuperates HMP Strange Loop #1; but the notion that that loop entails "an invasion of the body and its power in language" shifts Loop #1 so radically that we're inclined to envision HMP Strange Loop #2:

> By *the voice*, I mean orality. But no longer in the sense of the sign, where all we hear is sound opposed to meaning. In the continuum, orality is of the body-in-language. It is the subject we hear. The voice is of the subject passing from subject to subject. The voice makes the subject. Makes you subject. The subject makes itself in and through its voice.
>
> (136)

The voice as orality as "the subject we hear" is for the HMP not a phenomenology or psychology of reading but something actually *in* the text, or emerging from the text, which is, they say, "what a subject does to its language" (139). And, indeed, with this virtuous circularity we do now have a full-fledged HMP Strange Loop #2: the *voice* as subject is a text-effect that in effect says "I" to the reader as its "you," and so, as in HMP Strange Loop #1, subjectivizes the reader as someone who in turn (or perhaps simultaneously) subjectivizes the text. Circular intersubjectivization as omnidirectional knowledge-transfer. Where HMP Strange Loop #1 channels that circular intersubjectivization through the whole-body seriality of rhythm, HMP Strange Loop #2 channels it through the *taamicizing* mouthability of voice: "What we hear in it is not what it says but what it does" (137). What the voice-as-text-effect does is to turn us into the subjects that hear it and subjectivize it.

Rhythm for the HMP involves action, affect, and power, in a set temporal series or sequence that they often call "movement," and has the effect of regulating and intensifying meaning. Often enough the HMP's remarks on this head are not specified for either the rhythmicizations of Strange Loop #1 or the vocalizations of Strange Loop #2:

> In other words, more than what a text says, it is what a text does that must be translated; more than the meaning, its power, its affect.
>
> (69)

> It is in the inseparation of affect and concept that meaning finds its power and invention.
>
> (69)

> Meaning depends on the movement of meaning.
>
> (120)

> The problem is a poetic problem, in the sense where in order to hear and make heard the action and the power of speech, and not only the meaning of what is said, we must trace back the serial nature of the entire text, the sequence of the all-rhythm. Power yields meaning. Meaning, without power, is the ghost of language.
>
> (136)

Indeed, by aggregating both strange loops we can characterize the HMP's account of "poetry" as (1) embodied discourse that is (2) serialized in time, which (3) readers experience kinesthetically as (4) the movement of other people's bodies, which (5) tend phenomenologically to blur together with our own. (This is a reformulation of steps A-B-C-D, above: the seriality of C remains at 2, but the bodily-becoming-mental experience of D is serialized as 1-3-4, and the phenomenological blurring of experiencing "selves" in 5 is the temporal sequencing of B's intersubjectivity.) Let's unpack those:

1. The HMP follow the Emile Benveniste Pandemonium in calling this kind of language "discourse," but it is not the disembodied discursivity of poststructuralist theory, which, as the Brian Massumi Assemblage (2002: 2) put it, may "make and unmake sense" but does not *sense*. It is discourse as fully human speaking, as fully embodied conversation with an audience. For the HMP the subject of the poem may not be physiologically embodied, but they are engaged in a performative scene that is in every way experienced through

the body (D). For the HMP both bodily rhythms and vocal prosodies are part of what the Chaim Perelman Pandemonium (CPP: Perelman 1982: 36–40; see also Perelman/Olbrechts-Tyteca 1969: 144–8) call "techniques of presence," the use of body language to stage meaning for an audience: to draw their attention to the things the speaker/writer consider most important, away from the things that are of secondary importance (see Robinson 2016b: 130–3 for discussion). The HMP (Boulanger 2011: 75) say that rhythm "dictates gesturing," and gestures are for the CPP part of the rhetor's presencing repertoire, along with facial expression, posture, and vocalization/tonalization; the notion that rhythm *comes first*, conditions the other presencing instrumentalities, probably has a lot to do with the fact that the HMP's source language, biblical Hebrew, is an ancient language that mostly survives in written form. (The HMP found the Zionist resuscitation of biblical Hebrew as a spoken as well as written language in Israel as useful a guide to the biblical source text as "current French" was to their target text—which is to say, of no use at all.)

2. As in C, rhythm "serializes" the knowledge-translation. What a text does, it does in time: "In order to hear and make heard the action and the power of speech, and not only the meaning of what is said, we must trace back the serial nature of the entire text, the sequence of the all-rhythm." What is "serial" about a text, of course, is not the text itself, but the activity of reading it one word at a time, the reader's eyes moving steadily across the line of print—and again, in the performative scene the HMP reconstructs, the *mikra* or reading-aloud of the Hebrew Bible by a reciter, that steady movement of eyes across the line of print being converted into a steady stream of rhythmic spoken (or, according to Buber [Robinson 1993b], shouted) words. The act of reading, silently or out loud, serializes the text as spatial artifact into a temporal sequence. And it is that sequence, always emerging out of the engagement of a reader with a text, that generates subjectivity—both the reader as reading subject and the writer as writing-becoming-read subject.

3. Since for the HMP this (B/2) serial/sequential emerging of rhythm-as-subjectivity-as-knowledge is part of (D/1) the *body*-in-language continuum, we might identify it as a *sense of kinesis* (movement) or kinesthesia—an experience that is had by living bodies. Gestures are kinetic, body movements as body language, but when we *sense* our own gestures they are kinesthetic; and presumably, since we can't see the gestures dictated by rhythm in a text, they are something we sense as well, another kinesthesia. And while technically the "orality" or "mouthability" of HMP Strange Loop #2's voice is an enteroceptive

sense (perception of a body organ's functioning from the inside), the fact that the voicing of a spoken text moves through the mouth throat to lips in time puts it in the serial movement of HMP Strange Loop #1's rhythms as well, yielding four (or five, depending on whether we consider rhythm and the movement of speech to be the same thing or two different things) important kinesthetic experiences that would appear for the HMP to be constitutive of poetic subjectivity: (a) rhythm, (b) the movement of speech, (c) the movement of meaning, (d) gestures, and (e) orality/mouthability.

4. (a) Kinesthesia is traditionally the sense of *one's own body moving*, and the kinesthetic experiences in (3a-e) involve a sense of *other people's bodies moving*—or, perhaps, of textuality as someone else's mouthably rhythmic movement of speech and meaning that dictates gestures.

(b) But since the HMP explicitly theorize subjectivity as "the pursuit of a subject striving to constitute itself through its activity, but where the activity of the subject is the activity by which another subject constitutes itself" (35), we should reframe (4a) as *another's-becoming-one's-own* mouthably rhythmic movement of speech and meaning that dictates gestures (again, refer back to section 1.3). (c) And since, up until the moment we pick up the source text and begin reading it, it is black marks on the page, nothing living, we should perhaps further add that (4ab) is *first* (i) our own mouthably rhythmic/constitutive movement of speech and meaning; then, by the contagion of one subject's activity constituting another's, second (ii) the "subject of the poem"; then, finally, in a reconstituted "chronological" sequence, (iii) the poetic subject's mouthably rhythmic movement of speech and meaning becoming our own.

Observations (1-4) about rhythm would thus stand as a kinesthetics of knowledge-translation as an aggregated HMP Strange Loop #1/2. And let us remember that all talk so far of "texts" and "reading" and "writing" is HMP shorthand for a much broader scope of knowledge-translation, including the serial and intersubjective kinesthetics of nontextual and nonverbal learning in Communities of Practice—all social encounters in which embodied actions (5) tend phenomenologically to blur together with our own.

5. (a) One of the HMP's oft-repeated dicta is that rhythm conditions meaning; but they do not stop there. There is no direct line of force from the kinesthetics of rhythm to the cognitive formation of meaning or knowledge; the subjectivizing force they theorize moves through affect and power as well. If we think of affect as the full range of emotions, beliefs, doubts, moods, and so on traditionally

associated with it, and of power as conation—the power to move us to act—we get something like a kinesthetic-becoming-affective-becoming-conative-becoming-cognitive entelechy. This, obviously, might stand as an expanded and enhanced version of the affective-becoming-cognitive entelechy advanced in Robinson (1991) and section 4.1.

(b) Also as in section 4.1, the Aristotelian term "entelechy" implies not end-directedness, as the term is sometimes misunderstood, but "having an end within." The key word in that sequence in 5a, therefore, is "-becoming-." Each separate item in the sequence is constantly in the process of becoming the next, even if in some actual cases it doesn't click into place, as it were. Performative subjectivity as this sort of complex entelechial becoming may not become cognitive, for example: conscious, analytical, self-aware. Many writers are not aware of their own subjectivity in writing, and many translators are not aware of their own in translating. Many writers and rewriters are not aware of writing for an audience. But there is still an entelechial *movement* toward becoming self-aware.

(c) By the same token, this (5ab) articulated entelechial subjectivity may well be described as ethical even if no human being occupying one of the subject-positions is capable of conscious ethical choice. The fact that the entelechy is forever *becoming-conative* renders it becoming-ethical. Conation is directed or guided pressure to act, and thus participates in the social ecology of becoming-communal and becoming-normative (called "ecosis" or "ethecosis" in Robinson 2016b).

(d) Finally, this (5a–c) performative construction of subjectivity would not work without the circulation of kinesthetic senses, affects, conations, and cognitions through the group, from self to other(s) and from other(s) to self, mine-becoming-yours-becoming-mine, his-becoming-hers-becoming-mine-becoming-yours, and so on: "my"/our kinesthetic awareness of y'all's body language (including the rhythms in y'all's voice) generating not only affects and conations in "me"/us but a sympathetic sense of y'all's subjectivity as well; "my"/our tendency to experience y'all's affect conatively, as pressure to conform to it, and to the group norms it represents. The Maurice Merleau-Ponty Pandemonium (MMPP; Merleau-Ponty 1945/1970: 352) give a fairly simple example of this circulation when they put a fifteen-month-old toddler's finger in their mouth and pretend to bite it, and the toddler-pandemonium open *their* mouth, imitatively:

The fact is that its own mouth and teeth, as it feels them from the inside, are immediately, for it, an apparatus to bite with, and my jaw, as the baby sees it from the outside, is immediately, for it, capable of the same intentions. "Biting" has immediately, for it, an intersubjective significance. It perceives its intentions in its body, and my body with its own, and thereby my intentions in its own body.

Something like this event, which we now know to emerge out of the functioning of the mirror neurons, might be taken as a model for the kind of intersubjective mouthability that the HMP seem to want to theorize—which we might now number HMP Strange Loop #3. The toddler in the MMPP story "perceives its intentions in its body, and my body with its own, and thereby my intentions in its own body": the ostensible "immediacy" of pretend finger-biting, which according to the MMPP "has *immediately*, for [the toddler], an intersubjective significance," is actually serial, and therefore a rhythmicizing subjectification and/or subjectifying rhythmicization, but the events are serialized so rapidly by the mirror-neuron system—in around 300 milliseconds—that we experience them as simultaneous.

More specifically, the culmination of that rethinking in (5d) is what we call an "icosis" (also called a "doxicosis" in Robinson 2016b[3]): if *ta eikota* or "the plausibilities" or "things that seem true" are a communal construct, then the social ecology by which opinions are collectively transmuted into (the appearance of) truth is the sociological baseline of all knowledge-translation. Because the serial events of that icotic socioecological circulation tend to occur apparently simultaneously, and because the icotic "locking-in" that results is managed by the affects of the autonomous nervous system, they mostly fly under the radar of conscious awareness. This bypasses the DHP's dictum that "I feel, though, that such a hope [namely, that *feeling* may be at work in all this] is confused, because it uses what I consider to be just another synonym for 'conscious'—namely, 'aware'—" (2007: 276), obviously, but it also does more: it offers a possible explanation for the DHP's inability to recognize the affective substratum in the becoming-cognitive that they theorize as the totality of "cognitive science."

While the affective-becoming-cognitive mouth-and-ear for rhythm is in an important sense an icotic construct, in fact—built up through long and intense and sensuous icotic interactions with other speakers of the language—it is specifically the kind of icotic construct that tests the validity or authenticity or "truthiness" of other icotic constructs, and thus a kind of icotic metaconstruct (see Robinson 2012a: 92–6 for discussion). Icosis is specifically the socioecological process through which a community transforms individual-becoming-collective opinions into what come to feel like ontologically reliable realities and truths; the body-becoming-mind's mouth-and-ear for rhythm is one of the semi-individualized quasicollective "controls" or regulatory monitors of any given utterance's proximity to icotic ideals for not only elegance and pith but actual truth, semantic and thus "ontological" reliability. (Oscar Wilde: "style is truth.")

And, need we add, one of the "truthy" or ostensibly ontological "realities" shaped and ultimately constituted by that kinesthetic-becoming-affective-becoming-conative(-becoming-cognitive) mouth-and-ear for rhythm is the self—the "I." We are suggesting that the HMP's three strange loops can account for what the DHP calls (but neglects to theorize) the "locking-in" of the "I."

What we've accomplished here may, however, be meretricious—may amount to nothing more than setting up "rhythm" and "knowledge-translation" as what Kenneth Burke would call "god-terms" for human communication. To the extent that the kind of communal (self-)regulation that we have called icosis is a kind of deep intersubjective ecology, it is the basis and prototype of, and organizing force behind, all knowledge-translation; and the fairly narrow type of knowledge-translation that involves finding equivalents in language B for phrases in language A—or, for that matter, that helps novice writers find equivalents in their first language for the pandemonial thoughts coursing through their heads—would only be a special case of that. And to the extent that we follow the HMP in taking rhythm-and-voice as the default channel of all communication, and thus in our terms of all icosis, the specific problems of finding rhythms-and-voices in language B that align somehow sequentially with the rhythms-and-voices in a specific text in language A come to seem a rather paltry example of a process in which all of us are engaged virtually all the time (even first-year writers).

Still, we suggest that what is at stake here is rather more than the "hermeneutical" reduction that the HMP complain of in Heidegger and his followers, "this generalized essentialization of language, or poetry, of germanity that put the whole of translating into understanding" (Boulanger 2011:35). We are emphatically not putting the whole of knowledge-translating into icosis as social regulation. Rather, we are suggesting that thinking knowledge-transfer through the strange loops of rhythm, voice, affect, multiple intertwined embodiments, and subjectivization, and thus through embodied intersubjectivities, helps us recognize the ways in which it is more than communication—more, say, than an electronic mailing list. Thinking it as knowledge-*translation* also helps us put affective models of interlingual translation at the very explanatory core not only of knowledge management studies but of the sociology of human behavior in groups.

4.3 The Strange Loops of the Translator's Constructivist Agency: Kobus Marais

In section 4.2 we constructed "construction"—the social construction of "the affective-becoming-cognitive mouth-and-ear for rhythm," and more generally of reality, truth, and identity—icotically. Entailed in that construction is the assumption that icotic theory or icotic analysis is a social constructivism. While we feel quite comfortable with that entailment, social constructivism has recently come under rather passionate fire in translation studies, for reasons that we take to be partly good (no one, not even the social constructivists, seems to have any idea how social reality is constructed) and partly bad (phantasmatic claims projected out of fairly widespread ignorance onto social constructivism). The South African KMP, for example, noting that Mona Baker's *Translation and Conflict* (2006) "could be described as one of strong constructivism" (Marais 2014: 92), explain:

> If there is no shared reality outside of my consciousness and/or culture and/or narrative, then, yes, we have only conflict. In my explanation of complexity theory, I have, however, indicated that we live in one world, a physical-chemical-biological-psychological-social world. If one separates the psychological world from the social or physical world, it becomes easy to think in terms of constructivism. However, if the atoms in your body and in mine are kept together by the same physical forces, if we share the same chemical substances, if we share 99 percent of the same genes, is it so easy to claim that [a] we each construct our own world? Do we not have to take cognizance of the fact that [b] we are also constructed? As I understand it, this is the claim of sociologists, that is, that social reality constrains (constructs?) human possibilities just as physical reality does it. Also, if we consider the social and its constraints on our being, [c] are we not also socially constructed? Furthermore, [d] is constructivism not

a particularly Western philosophy, and does it not arise from an ideological stance where [e] people think that they have conquered nature? [f] Are people living in abject poverty or extreme heat or cold able to hold to constructivist theories? I am well aware that constructivism is, in part, a response to essentialist tendencies in social thought. However, denouncing the essentialism or natural determinism in social relationships does not have to lead to the other extreme, strong constructivism. I am thus questioning the [g] radical forms of constructivism that are currently rife in translation studies. A more ecologically sensitive model ... is more tuned to the complexity of the human condition.

(92; bracketed letters added)

The ground-zero inaccurate guess at the claims made by constructivists there is that (a) "we each construct our own world"—that because constructivists see social reality as constructed, it must be a *conscious choice* made by individuals, and made *differently* by different individuals. That's wrong. Nobody thinks that.

The KMP's next inferential assumption, as their imagination works on this phantasmatic a-construction of constructivism, is that (d) it's a "particularly Western philosophy" that arises from (e) "an ideological stance where people think that they have conquered nature." After all, if "strong" constructivists believe that (a) each of us constructs our own reality, surely that must be an egregious colonial fantasy of agentive omnipotence. Those "critical translation theorists" project onto translators not only fully conscious decision-making but Heroic Agency, Super-Agency, King Reason at the helm of the Free White European Male, the power to shape reality in each individual's personal image, and that is not only wrong but an expression of Western power and privilege. No need to refute that: it's just an imaginative working-out of the inaccurate a-guess at ground zero.

One step beyond that (a-d-e) is the claim that this (g) "radical constructivism" is "currently rife in translation studies." It isn't some weird minority view: it's "rife." It's everywhere. It pervades the "critical theory" of translation. This is not an inference like d-e; it's a fantasized truth-claim like a.

And, of course, once that paranoid image has taken hold, the KMP go to the *reductio ad absurdum*: (f) "Are people living in abject poverty or extreme heat or cold able to hold to constructivist theories?" Presumably what they mean by that is not whether "people living in abject poverty or extreme heat or cold" are able to read and understand and accept high academic theory, but whether they might be inclined to believe in their own sovereign power to create reality *ex nihilo*, and thus in their ability to control their environments through the sheer force of their will. "In a postcolonial context," they note elsewhere, "it is an open question whether people have that power" (66). Only a European or North American, the KMP snort, would harbor such delusions. Africans, never having had that power, are humbler, and so less susceptible to such *ignes fatui*.

Needless to say, this kind of conscious rational control of social reality is not what social constructivists mean by the social construction of reality. It may perhaps be needful to reiterate and underscore as well that no translation theorists are making claims like this. The whole line of attack is a social "reality" constructed entirely in the KMP's imagination—by King Reason, perhaps, at the helm of the Free White African Male. Given that it is a paranoid fantasy—using "paranoid" not in the clinical sense, to diagnose the KMP demons as psychotic, but in the ESP's sense, as "one kind of epistemological practice among other, alternative, ones" (Sedgwick 2003: 128)—one might be inclined to dismiss it out of hand, as an embarrassing but not particularly useful error. The invocation of the ESP's brilliant chapter on paranoid and reparative reading in *Touching Feeling*, however, suggests that the very embarrassment this "kind of epistemological practice" evokes may point us in very useful directions. The ESP's book, after all, is affect theory: perhaps the KMP's plunge down the rabbit hole might, followed closely enough, lead us back to bodies, with new insights?

There is one page in the KMP's book, in fact—144—where they continue to complain about the colonial delusions of the West, but also tentatively sketch out a possible middle ground between (c) the elitist super-agency/constructivism of Western critical theorists of translation and the utter lack of agency/constructivism that they project onto (d) "people living in abject poverty or extreme heat or cold." It's not quite clear what that middle ground is—we see the KMP throwing up their collective hands and professing not only their own but everyone else's uncertainty about agency in general—but it does seem to involve a *humble* and perhaps *passive* kind of constructivism, such as is hinted at perhaps in the passive verbs in the rhetorical questions above: "Do we not have to take cognizance of the fact that *we are also constructed*?" and "*are we not also socially constructed*?" That is, in fact, far closer to how social constructivists actually believe social reality is constructed; more on that in a few pages.

With y'all's indulgence, we'd like to quote almost the entire page, as to our minds it is the most interesting and significant page in the entire book:

For now, [1] I suggest that translation scholars rather work empirically to trace and account for the interfaces of translational and developmental actions. [2] I do not suggest a total ban on judgment and criticism, but instead a temporary moratorium, until we know and understand enough. I would not want my work to be suggesting yet another crusade to save the world.

Studying development from a translation studies perspective will require more thinking on agency. [3] The type of activist agency currently advocated in translation studies will have to be revisited. [4] As (a part of) translation studies frees itself from its bondage to critical theory and academic activism, it will be able to see that perhaps there are many other ways of being agents than being activists. [5] Consider the millions of pages of translated text that go into the construction of the European Union. Also consider the billions of translated spoken words that make the running of the informal economy in developing countries possible. Further consider the translation of ideas on economic reform from Washington to Kampala or on plant breeding from the University of the Free State to a rural farm in Ethiopia and its role in developing the latter. [6] The kind of agency involved in these actions needs to be thought about, and to refer to Latour (2007), agency is the one thing we know virtually nothing about. [7] Thus, for now, I am not judging a particular theory of development to be better than others. I am not suggesting a particular way of developing through translation. [8] My aim is to understand and account for what people do when they need to communicate, in whichever mode, with other people who do not speak the same language as they do. [9] Furthermore, my aim is to understand and account for the ways in which these communications contribute to the creation and maintenance of social ties or "the social".

[10] I thus contend that the focus on agency in translation studies is part of a Western analysis of reality. [11] You can only contribute if you are actively for or against something. [12] It also rests on a very strong belief that your actions matter and that you are in control of history and nature, that is, humanism. [13] Nonlinear systems theory relativizes the importance of human agency. The outcome of your input cannot necessarily be predicted.

I am not arguing that one should forego the notion of agency. [14] What I suggest is that we look for other modes of agency, that is, translation that serves or translation that builds. These forms of motivation for action are also agentive in nature. [15] What I am trying to say is that agency in the critical theory definition of the word is not necessarily the only kind of agency contributing to the construction of social reality. [16] The typical anonymous, voiceless, invisible translator slaving away in a stuffy little office, translating boring municipal regulation after regulation, is contributing as much if not more to the construction of social reality than the verbose literary translator who performs an aggressive feminine translation of a literary classic. [17] Western notions of high visibility, branding, and status should not be the only ones defining the agency of translators. (2014: 144; bracketed numbering added)

By way of getting started with this, some observations. Three, to be exact: (A) the problematic claim (again) about "strong constructivism" among "critical theorists" of translation; (B) the problematic nature of the "humble" or "passive" constructivism that the KMP seem to be imagining as an alternative to "strong constructivism"; and (C) the KMP's reductivist binarization of "constructivism" and "complexity theory."

A. It is again quite obvious that the KMP's conception of "agency in the ['activist'] critical theory definition of the word" (3–4, 10–12, 15, 17) is their own pandemonial invention. It's not a straw-man exaggeration; it's a fantasy. The KMP never give us a single textual illustration of the supposed constructivist attitudes they are attacking—presumably because such texts do not exist. They never openly engage the theorists who supposedly exemplify this "Western" constructivism, by name or by text. They just attack, eyes squinched tight shut. Those horrible critical theorists (among whom, full disclosure, we count ourselves) are invoked purely and anonymously as whipping boys, girls, and others. A generous reading would speculate that the KMP demons have never read anything about social constructivism, never made the slightest attempt to understand its basic tenets, but rather are making guesses, based on their vague impressions of what the critical theorists they despise must be implying.

If there are in fact critical theorists who believe that (11) "You can only contribute if you are actively for or against something," we'd like to know about it. It would have been helpful if the KMP had listed some, and quoted their telltale pronouncements to that effect. There are, to be sure, translation scholars who *study* activist translation, especially in politically polarized contexts and conflict situations where all human beings, translators or not, are required to take an ethical stand (or else hide); there are even some translation scholars who advocate for activist translation. But none of them seek to generalize that kind of activism to all translators in all possible contexts, let alone to prescribe activist choice as some kind of universal model of "translator agency." That's the fantasy.

The claim that translators (12) "are in control of history and nature" is another instance of the KMP's inferential chain from a to d to e, above. Nor have we seen the tiniest shred of textual evidence, even in conversations, conference papers, and rejected article submissions, that (17) "Western notions of high visibility, branding, and status [are anywhere, in anyone's but the KMP's account] defining the agency of translators." What we do see are studies of the translator's "narratorial" style—of tiny hints that the translator is a human being with a personality, hints so tiny that they tend to be isolated and speculatively identified as indicators of a personal style through the analyses of large corpora. That's not exactly "high visibility, branding, and status."

To be sure, the KMP's snide attack on the edgiest of feminist translation studies—(16) "the verbose literary translator who performs an aggressive feminine translation of a literary classic"—suggests that they've been reading Quebecoise feminist translators and translation scholars like Lotbinière-Harwood (1991). If so, it would have made for a stronger case had they cited Lotbinière-Harwood and generalized persuasively from those quotations. The problem with even that more reasoned approach, however, is that generalization from radical feminist translation practices and strategies, or even from radical intervenient translation practices and strategies more broadly (see Munday 2007), to "critical translation theory" as a whole is a non-starter. The only way an attack on radical intervenient translation can be construed as supporting the attack on an imagined heroic conception of the translator's agency supposedly perpetrated by Western colonizing critical theorists is if one assumes (a) that feminist translation scholars like Lotbinière-Harwood are universalizing their declaration of intervenient feminist translation to all translation (and they aren't); (b) that feminist intervenient translations are emblematic of all critical theorists' understanding of the translator's agency (they aren't either); and especially (c), that intervenient translations are necessarily "aggressive" (they can be playful, creative, funny, coy, quirky, self-reflexive, even self-undermining).

Suppose we were to take Susanne de Lotbinière-Harwood's feminist anger at male domination, and determination either not to translate male writers at all or to intervene translationally in a male source author's message, to be both "verbose" and "aggressive." To someone who considered accurate translation and male prerogatives to be sacrosanct, that "aggression" would be a bit scary and shocking, to be sure. Someone like that might be tempted to lash back viciously: how dare you! That kind of fear- and taboo-driven backlash is unpleasant, but perhaps psychologically understandable. Privileged men—even, incredible as this may sound, white Afrikaners—do not like the tiniest threat of diminished status and power. But it's sheer demagoguery to turn that kind of anxious irritation into generalized fear-mongering aimed at a whole class of people. (Not that that kind of demagoguery is exactly a rarity among privileged pandemonia in white cis-straight-male bodies.)

B. The KMP's alternatives to this fantasy of Western celebrations of the translator's godlike control of history and nature are equally vague and shadowy—and sometimes quite disturbing. The mythic liberation narrative of (4) "As (a part of) translation studies frees itself from its bondage to critical theory and academic activism, it will be able to see that perhaps there are many other ways of being agents than being activists" seems to promise a glorious world of freedom beyond the prison walls—but, of course, freedom is precisely what the KMP attributes to the carceral "bondage to critical theory and academic activism" that they decry, so what we are explicitly offered is either (5) nondescript to the point of opacity or aggressively antiheroic: (16) "The typical anonymous, voiceless, invisible translator slaving away in a stuffy little office, translating boring municipal regulation after regulation." It may be difficult to guess what the KMP is implicating in (5) "the construction of the European Union," "the running of the informal economy in developing countries," and so on, but at least guessing is possible. The problem with that (16) "typical anonymous, voiceless, invisible translator slaving away in a stuffy little office" is exactly the opposite: the KMP's depiction of that slavish translator is not just a savage attack, and not just an attack whose savagery rivals that of the KMP's attack on "activist" "critical theorists," but an aggressively *explicit* attack. They make no attempt to paint this "real" translator as a complex human being with mixed motivations, a love-hate relationship with the work, or the like; rather, every attribute is carefully chosen to heap scorn on this person. They are not just "anonymous, voiceless, [and] invisible"; they are not just "slaving away in a stuffy little office": the municipal regulations they are translating are *boring*. There is not a single redeeming quality to this "real-life" alternative to the godlike devil-Westerner from whose constructivist clutches the KMP seek to rescue translation studies. They are another vicious stereotype, out of some Gogol or Dostoevsky short story.

And of course the alternatives presented to us by the KMP are imagistically so similar, and so similarly dystopian—either "*bondage* to critical theory and academic activism" or "*slaving* away"—that we seem to be in a no-exit nightmare.

Interestingly, however, the KMP also construct that translator-slave as "contributing as much ... to the construction of social reality"—possibly even more than the nasty aggressive verbose feminist translator! How is that possible? What is this "humble" or "passive" or even "slavish" constructivism that the KMP's account seems to adumbrate? How does it work? The use of the noun "construction" in (5) "Consider the millions of pages of translated text that go into the construction of the European Union" seems to be hinting that those EU translators are also working to construct the historical reality of the EU as well—and, of course, they are. But there would seem to be a significant difference between "slaving away" translating municipal or EU documents, and in that sense helping to maintain a social institution, and constructing *reality*. The KMP's account of EU translators (and probably slavish translators of municipal regulations as well) seems to point to the former, namely institutional maintenance; their attack on the "strong constructivism" or "radical constructivism" of "critical theorists" among translation scholars seems to point to the latter, namely the Heroic Super-Agency of world-dominating Westerners.

And it may be that the intensity and the apparently gratuitous viciousness of the latter attack has something to do with the KMP's inability to imagine the notion that *reality*—including sensory reality—is socially constructed. Kantian idealism and its later post-Kantian developments, especially perhaps Lev Vygotsky's social constructivism, would appear to be simply incomprehensible to the KMP, who confessed to us that their stubborn and combative realism has a lot to do with growing up on a farm in the Groot Karoo in the Western Cape Province of South Africa. One can't "construct" the weather, they told us.

That, of course, is a simple misunderstanding of what is meant by "construct" in post-Kantian social constructivism; we'll return to discuss a series of correctives to that in a moment. The significance of the misunderstanding for now is that it explains the map that the KMP seem to be attempting to chart in these two passages. It makes it very clear that the apparent middle ground the KMP seem to be hinting at between (a-d-e) the elitist super-agency/constructivism of Western critical theorists of translation and (g) the utter lack of agency/constructivism that they project onto "people living in abject poverty or extreme heat or cold" is not a middle ground after all. It's a new binary, between helping to maintain social institutions (even as slavish functionaries like municipal or EU translators, even perhaps as farmers trying to maintain the viability of their farms despite their inability to control the weather) and the colonialist vision of controlling nature and history. For the KMP the former is good because realistic; the latter is bad because absurdly unrealistic. The functionaries and the farmers may not be High Romantic heroes, in fact they may seem slavish and boring, but they're doing honest work that helps maintain society. The High Romantic heroes that the "constructivist" "critical theorists" seem to want translators to be are a mere silly pipe dream, and their overweening arrogance is a sham.

C. This unstated binary between maintaining institutions and creating and controlling reality that we speculatively assign to the KMP also explains, we suggest, why they are so confident that "constructivism"—at least its "strong" or "radical" form—is utterly at odds with their adherence to complexity theory. Let's look again at the first few lines of the first quotation:

> If there is no shared reality outside of my consciousness and/or culture and/or narrative, then, yes, we have only conflict. In my explanation of complexity theory, I have, however, indicated that we live in one world, a physical-chemical-biological-psychological-social world. If one separates the psychological world from the social or physical world, it becomes easy to think in terms of constructivism. However, if the atoms in your body and in mine are kept together by the same physical forces, if we share the same chemical substances, if we share 99 percent of the same genes, is it so easy to claim that we each construct our own world?
>
> (2014: 92)

Reading this, anyone who knows the theory of social constructivism is scratching their head: why would anyone assume that *social* constructivism posits "no shared reality outside of my consciousness and/or culture and/or narrative"? The whole point is that the constructed reality outside any individual's consciousness is *social*, *socially* constructed, and therefore shared. The solipsism that the KMP project onto social constructivism is a big part of the problem in their account. Social constructivists would in fact agree with the KMP that "we live in one world, a physical-chemical-biological-psychological-social world"—but only to the extent that "we" are defined locally, as the members of a specific culture. *We here in this local culture* all do live in the same world, because we collectively construct it. In terms of the KMP's cultural context, which we're just now leaving, as we write this—sitting in Johannesburg, waiting for our flight home—we Afrikaners live in one world, and those Xhosa live in another, and those Zulus live in yet a third, and those Basothos over there live in a fourth, because each culture constructs (imagines, understands) their reality in a different but internally organized way. Those distant "Westerners" in their countries up north live in a totally different reality, or collection of different realities.

It's possible, of course, to dismiss those cultural differences and insist that there is only one "scientific" or "empirical" reality, and anyone who for whatever perverse reason deviates from that is simply "primitive" or whatever. It's possible, in other words, to universalize one's own sociocultural construction and despise all other groups. This is precisely the colonial arrogance that the KMP so passionately critique, but also, perhaps, surreptitiously exemplify. The inclination to naturalize one's own sociocultural construct as the only "empirically" true one is ethnocentrism, obviously, but also in this case a survival of naive pre-Kantian scientism, objectivism, the ideologically normative Western belief that it is possible to know objective reality reliably. We assume, in fact, that the KMP's rejection of social constructivism is strongly fueled by this kind of objectivism: for them it's simply absurd to believe that one can control the weather.

Tell that to the shaman leading the rain dance. That shaman's absurd, right? Absurd to anyone outside of that cultural context—especially to a Westerner. Absurd to a Groot Karoo Afrikaner farm-boy like the KMP as well, I'm guessing. But then the novel we bought in Bloemfontein by Sally Andrew, *Death on the Limpopo* (2019), has a main character whose Klein Karoo Afrikaner culture seems to believe that everything is alive. Regional difference, with a mountain range in between? Gender difference? But never mind that: let's have a good laugh at everyone in the world who doesn't share our sociocultural constructs of reality. After all, that kind of ridicule of the other helps us consolidate our own group's sense of cultural belonging, and thereby also sense of the intrinsic *rightness* of our beliefs about the universe.

Where are the strange loops here? Well, the problem of social constructivism is the problem the DHP face in explaining how we "lock in" to the phenomenology by which, after several years of an infant's and toddler's strangely loopy interactions with their environment, "pretty soon this high abstraction behind the scenes ['I'] comes to feel like the maximally real entity in the universe" (2007: 179). Somehow, "via the loop of symbols sparking actions and repercussions triggering symbols, the abstract structure serving us as our innermost essence evolves slowly but surely, and in so doing it locks itself ever more rigidly into our mind" (186). But how does an "abstract structure" *evolve*? What is this "locking-in"? The DHP don't know. All they can do is model it technologically: "Indeed, as the years pass, the 'I' converges and stabilizes itself just as inevitably as the screech of an audio feedback loop inevitably zeroes in and stabilizes itself as the system's natural resonance frequency" (186). It's an inevitability. It doesn't matter *how* it happens; it's enough to note that audio feedback loops also stabilize themselves. If an audio feedback loop "locks in," why shouldn't the strange loops that construct a toddler's "I" do the same? After all, what possible difference could there be between a toddler and a sound-reproduction device?

The same set of questions applied to the "reality outside of my consciousness and/or culture and/or narrative" would address the problematic that to our knowledge has not yet been addressed in social-constructivist theory, namely how it's possible that the sensory reality in which we live "comes to feel like the maximally real entity in the universe." Surely the chair we're sitting in, the desk our computer rests on, the coffee cup beside the computer on the right and the smart phone beside the computer on the left, and the airplanes and tarmac personnel we can see across the freeway through our hotel window are all *real entities*? Surely we're not just imagining them, "constructing" them? If, according to the KMP, "we live in one world, a physical-chemical-biological-psychological-social world," perhaps there's some variation in the psychological and social emergences in that sequence, but surely it's the same physical-chemical-biological world? The *felt reality* of the physical world seems like an irrefutable proof of the rightness of objectivism. It is tangible. If we rap our knuckles on the desk, carefully avoiding tipping over the coffee cup, our knuckles hurt. Proof, right? The desk *really exists*. It's a stable object in the world. It's not a social construct.

And, sure, we know that our senses are a bit unreliable. We know that our nervous system distorts the sense-data en route to the brain. We know about phantom pain in amputated limbs. We know that congenital analgesiacs don't feel the pain in their knuckles when they rap them on a desk, or any other pain either. They can run their hands over the desk with their eyes closed and confirm that it is indeed a desk, but they don't feel pain when they rap their knuckles on its surface. We know, if we're reasonably well read in the cognitive neuroscience of the individual nervous system, that most of that can be explained along (post-)Kantian lines: an "objective" reality exists, but we have no "objective" access to it. All of our experience of that reality is mediated for us by our nervous systems, and our nervous systems are unreliable. Congenital sensory neuropaths are born with the ability to feel pain and sense hot and cold, but lose it in early infancy. Their discriminatory touch—the ability to distinguish a desk from a kitchen counter, say—is often impaired as well. The "immediate" sensory experience that normals have of objective reality is a neural construct. It is not immediate at all. It only feels that way, thanks to the normal mediatory functioning of our nervous systems.

But that's the *individual* nervous system. What confuses the DHP in theorizing "locking in" is that they fail to make the connection between the toddler's experience of the "I" locking in and the Doug-Carol experience of sharing pandemonial personalities. The view from the social neuroscience of the simulation of other people's (and pets', etc.) body states by the mirror neurons is quite different. Not only is it possible to feel someone else's pain, somatomimetically; it's possible to feel that someone's somatic markers guiding attitudes and actions regarding that pain. Whatever one human pandemonium has learned through experience, to the extent that those lessons have been stored in the autonomic nervous system and activated as behavioral guidance by somatic markers, can be simulated by another human pandemonium's mirror-neuron system, so that the pandemonium who didn't directly learn those lessons can internalize them as behavioral guidance as if they had learned them directly.

This mimetic *sharing* of somatic markers is the key to the DRP's theory of "locking in." It's not just that the individual autonomic nervous system stores experiential lessons and reminds the organism of them through somatic markers; it's that somatic markers are *passed around* through a group, circulated, as in section 4.2 (pp. 155-8). This somatomimetic mirroring/simulating of autonomically stored action-orientations across bodily boundaries serves to *normativize* the action-orientations of which the group approves and to denormativize, even "demonize"—in the Christian rather than the pandemonial sense—those of which the group does not approve. Approval and disapproval, as social affects, are somatic markers that can be reticulated through a group at such high speeds—within 300 milliseconds—that they seem instantaneous.

Hence the *social* construction of reality—including the social construction of the "I," through the locking-in of analogical patterns triggered through strange loops. It doesn't happen automatically, mechanically, "inevitably," as in an audio or video feedback loop; it gradually comes to be stored as somatic markers in both the infant and toddler and their caretakers and other companions. And because groups differ, because the icotic construction of reality is managed locally, in cultures that have their own norms and values and inculcate them in their children somatomimetically as icotic realities, "reality" "itself" can look very different depending on what group(s) one belongs to.[4]

This social-neuroscience view casts a very different light on the KMP's observation that "In my explanation of complexity theory, I have, however, indicated that we live in one world, a physical-chemical-biological-psychological-social world. If one separates the psychological world from the social or physical world, it becomes easy to think in terms of constructivism." For the KMP's "explanation of complexity theory" in Part I of the book (Marais 2014), living in "one world" means that there are emergent continuities from one level to the next, so that the psychological emerges out of the biological and the social emerges out of the psychological; but the warning against "separating the psychological world from the social or physical world" suggests that for the KMP emergence is never separative—can *never* involve any kind of separation. As we understand complexity theory, this expectation of stable or at least predictable emergences is actually reductivist: the physical-chemical-biological substrate of the brain in that case *controls* the emergence of the psychological-sociological, making it possible to explain the psychological and the sociological on the basis of the physical-chemical-biological substrate. "One world" in this construction sounds very much like a closed system.

Post-Kantian thought, including Lev Vygotsky's social constructivism, is in fact far more compatible with complexity theory than the reductivist objectivism that the KMP seem to be hinting at here. Because we cannot *know* the physical-chemical-biological world reliably, the emergence of "the" psychological and sociological worlds out of it actually entails both a wild and unpredictable proliferation of emerging psychological-cultural-sociological worlds and a rampant downward causation whereby those emergent social worlds variably construct the physical-chemical-biological world out of which they emerge. The KMP's belief that the emergent social worlds *cannot* construct those "lower" worlds, indeed that it is some kind of Western colonial-overlord arrogance to think that they can, is reductivist objectivism: the physical-chemical-biological world for them *is what it is* no matter what those power-hungry Western constructivists might think.

The best psychological metaresearch we know on the social construction of reality is presented by Allan N. Schore, the great integrator of psychological, psychoanalytical, and psychobiological research on affect-regulation in our time, in a series of books entitled *Affect Regulation and the Origin of the Self* (1994), *Affect Regulation and the Repair of the Self* (2003a), and *Affect Dysregulation and Disorders of the Self* (2003b). Consolidating vast quantities of psychological research based on Bowlby's attachment theory, Kohut's self psychology, and Damasio's somatic-marker hypothesis, Schore tracks an emerging consensus that understands affect-regulation and dysregulation as products of nonverbal and largely unconscious/preconscious right-brain-to-right-brain communication—specifically, the communicative synchronization of affect through the mirroring of body language. Schore does not attempt to account for the expansion of dyadic right-brain-to-right-brain communication to larger social groupings; but it seems reasonable to assume that a group of two caretakers and an infant would also involve three-way regulatory communication, and that, as the social groups grow in size, we also learn to circulate regulatory communicative events—mimetic simulations of other bodies' somatic markers—across larger numbers of "dyads."

As Schore notes, "Mutual gaze interactions increase over the second and third quarters of the first year, and because they occur within the 'split second world of the mother and infant' (Stern 1977) are therefore not easily visible" (2003a: 8)—and therefore easily dismissed as not actually happening. Three decades ago Beebe and Lachmann (1988) studied those interactions by comparing simultaneous films of the mother and the infant frame by frame, finding "synchronous rapid movements and fast changes in affective expressions within the dyad" (Schore 2003a: 8). The Allan Schore Pandemonium (ASP), channeling thousands of researcher-demons, note that "this affective mirroring is accomplished by a moment-by-moment matching of affective direction in which both partners increase together their degree of engagement and facially expressed positive affect. The fact that the coordination of responses is so rapid suggests the existence of a bond of unconscious communication" (8). "Suggests"? Perhaps this is just researchers' imagination? Perhaps. Schore asserts that "these mirroring exchanges generate much more than over facial changes in the dyad; they represent a transformation of inner events" (8), but who knows? The ASP report that for Beebe and Lachmann (1988), at least, "as the mother and the infant match each other's temporal and affective patterns, each recreates an inner psychophysiological state similar to the partner's" (8). Stern (1983), the ASP note, found that "in synchronized gaze the dyad creates a mutual regulatory system of arousal," and in that state both mother and infant "experience a state transition as they move together from a state of neutral affect and arousal to one of heightened positive emotion and high arousal" (8).

The emerging consensus is apparently that in these "mutual gaze interactions" what is actually happening is that the infant pandemonium is imprinting the affective output (demons?) of the mother's right cortex, and in that way hard-wiring the circuits in their own right cortex for expanding affective capabilities. For the ASP the mother is "downloading programs" into the infant's brain (13), and thus serving as what Diamond, Krech, and Rosenweig (1963) call the child's "auxiliary cortex" (quoted in Schore 2003a: 13). The mother thus becomes the infant's "self-object," Heinz Kohut's (1968, 1971) term for the narcissistic state in which the other is taken as existing entirely in the service of the self; in later work (1977, 1983) Kohut dropped the hyphen and began to write it selfobject to indicate that the selfobject is not a pathology, an archaic fixation in which a grandiose self uses an omnipotent self-object to make itself feel bigger and stronger and more coherent, but rather a meaningful dimension of experience. For Taylor (1987) and Palombo (1992) the infant's selfobjects (beginning with the mother, in this research) become external regulators of the infant self's affective experience, and are gradually internalized as self-regulators; according to the ASP, "they act at nonverbal levels beneath conscious awareness to cocreate states of maximal cohesion and vitalization (Wolf, 1988)" (14).

Now we're not sure what the DHP would say about this model. There is to our mind nothing in it that is incompatible with the DHP's strange-loops theory; all we have to do to integrate the two would be to add *cognitive* (layered-analogical) strange loops to the affect-regulation model and *socioaffective regulation* to the strange-loops model. The cognitive strange loops would then explain where the cognitive idea of the "I" comes from, and the affect-regulation would explain how that idea "locks in." One would also, of course, need to add affect-regulation in larger and larger groups—the cultures of whole families, circles of friends, classrooms, workplaces, genders, races, social classes, language groups, education levels, nations, and so on, all the way up to the culture of the entire human race—in order to extend the DHP's image of "locking in" to social constructivism. But with these minor adjustments, the various models fit together snugly, and form in the end a coherent composite model.

Note, though, that we have one more wrinkle to account for, namely the "automatic" "inevitability" of the strange-loops constitution of the "I" in the DHP's theorizing. In the ASP's psychological metaresearch, "bad parenting" and other dysregulatory affect-regulation in the early years of an child's life can disrupt and even derail the constitution of the "I." On the one hand, "we know that the caregiver is not always attuned; indeed, developmental research shows frequent moments of misattunement in the dyad, ruptures of the attachment bond" (11), and in extreme cases of neglect children grow up with no attunement at all. On the other hand, we also know that "reciprocal gaze, in addition to transmitting attunement, can also act to transmit misattunement, as in shame experiences" (11), and in extreme case of abuse children grow up with various psychopathologies often aimed at destroying the selfobject projected onto other people. "The misattunement in shame, as in other negative affects," Schore continues, "represents a regulatory failure and is phenomenologically experienced as a discontinuity in what Winnicott (1958) called the child's need for 'going-on-being'" (10–11). As Schore describes these dysregulated states:

> Due to the impaired development of the right-cortical preconscious system that decodes emotional stimuli by actual felt emotional responses to stimuli, individuals with poor attachment histories display empathy disorders, the limited capacity to perceive the emotional states of others. An inability to read facial expressions leads to a misattribution of emotional states and a misinterpretation of the intentions of others. Thus, there are impairments in the processing of socioemotional information.
>
> In addition to this deficit in social cognition, the deficit in self-regulation is manifest in a limited capacity to modulate the intensity and duration of affects, especially biologically primitive affects like shame, rage, excitement, elation, disgust, panic-terror, and hopelessness-despair. Under stress such individuals experience not discrete and differentiated affects, but diffuse, undifferentiated, chaotic states accompanied by overwhelming somatic and visceral sensations. The poor capacity for what Fonagy and Target (1997) called "mentalization" leads to a restricted ability to reflect upon one's emotional states. Right-cortical dysfunction is specifically associated with alterations in body perception and disintegration of self-representation (Weinberg, 2000). Solms also described a mechanism by which disorganization of a damaged or developmentally deficient right hemisphere is associated with a "collapse of internalized representations of the external world" in which "the patient regresses from whole to part object relationships" (1996, p. 347), a hallmark of early forming personality disorders. (47)

That "disintegration of self-representation" suggests the partial or total failure of the process of "I"-constitution that for the DHP can usefully be modeled on audio and video feedback loops. Extreme forms of this failure would include ipseity disturbance or "self-disorder," in which the so-called minimal self—the feeling that one's experiences are one's own, minimal in the sense that it lacks the narrative content of the extended self—is severely diminished or disrupted. Because the process by which the minimal self is generated in humans is regulated affectively, through right-brain-to-right-brain communication across mutual gaze interactions, it doesn't just happen; it has to be nourished, cocreated, organized, and expanded socioaffectively-becoming-sociocognitively.

So how would that work, exactly? Let's start simple. Go back to my coffee cup on the desk. Imagine we have a toddler in the room with us, and the coffee is scalding. The toddler reaches up for the coffee cup, and—what happens? We meet their gaze, say "No," and pick up the cup, lift it down so they can look inside. "See?" we say. "Hot. Very hot. Don't touch." The toddler looks at the coffee, looks up at us, and thinks. We can see the little mind churning with this. What we can't feel is the toddler's affect—or even our own, for that matter, though if we think about it a little we've just had a little scare. Scalding hot coffee on this toddler's face would be a horrific event. Because we reacted quickly and appropriately, the scare is minimal—but it's still there. We would also assume that the toddler internalized that little scare—just a little, maybe.

But now rewind a little. Go back to before we filled the cup with hot coffee. Imagine the toddler grabbing the cup off the desk now. What is that toddler thinking? What do they know about cups? Rewind further: think of how a tiny infant learns what a cup is (mentally "constructs" the object as a cup): that it has an inside and an outside; that it has an up direction and a down direction; that one can put liquid in the cup, but if one does, the up direction has to be up, or *oops*. Think of that learning in terms of the DHP's strange loops: the hundreds of iterations of the visual/kinesthetic loop of seeing and feeling and trying the cup and noticing the seeing and feeling and trying of the cup. Think of it in terms of the ASP's affect-regulation: the hundreds of mutual gaze interactions with caregivers in which very simple messages like "safe, go ahead" and "dangerous, stop" are sent and received and processed. Now mentally combine the two models. They go together nicely, right?

Now jump ahead to the translator. Is there anything that the assorted demons of the Target Author Pandemonium (TAP) do that constitutes a similar kind of constructivist activity?

Well, we've seen several examples of this. In Chapter 2, the TAP construct the source text. It's just black marks on the page at first, and then a series of words and sentences and paragraphs; it has to be constructed as a text, and as a voice or series of voices, and as a series of rhythms, a vocal/rhythmic signature as it were. The TAP don't necessarily have to construct the Source Author Pandemonium (SAP), but we personally (the DRP) almost always do—especially when the SAP demons are alive and interacting with us on the translational choices we've made, obviously, but even when the SAP demons are dead or unknown. Somehow the construction of voices and rhythms just "naturally" seems to conduce to the construction of SAP personality, and even SAP body language. That constructivist process isn't literally natural, of course; but after four decades of translating, however, and of *loving* to translate, to us it feels natural. That feeling is something that we've "locked into," we suppose. The TAP tend to construct "a" or "the" Target Reader Pandemonium (TRP), too—someone or some multitude that they may never meet, and may never even exist. Still, constructing a TRP provides useful guidance in the choice of words, sentence structures, registers, and so on.

It should go without saying, too, that no one knows at birth, or even during early exposure to other languages, how to construct an SAP or a TRP, or oneself as a TAP—any more than we are born knowing how to use a cup. This ability is developed through practice, each tiny lesson as we learn it locked in through and as somatic response, those lessons felt again as signaled through somatic markers, those somatic markers as signaled swapped with the other pandemonia with whom we interact. And as the DHP would remind us, that socioaffective process is accompanied by a cognitive cycling around strange loops, a noticing (more or less or not at all consciously) of analogical parallels between figuring out what the source text means and becoming aware of oneself figuring out what the source text means; between guessing at what the SAP was trying to say and becoming aware that one is making guesses like that, and then noticing that one is using memories of guessing at what conversational interlocutors were trying to say by paying attention to their body language; and between guessing at how the TRP will respond to what one is writing and becoming aware that one is making guesses like that, and then noticing that one is using memories of guessing at how conversational interlocutors were interpreting what one was saying by paying attention to their body language. Around and around these strange loops—with some trepidation, at first (can we do this? can we be a translator?), but then with growing confidence, confidence bricked together somatically and therefore below the radar of our conscious awareness, but seconded by cognitive observations in the world: as one's clients are not unhappy, and perhaps even very happy, things must be going okay.

We personally remember the mixed feelings with which we received well-meaning advice from our second client, a professor of psychology that we knew at our university—an "older" person (they may have been in their forties or fifties, while we were in our very early twenties) who sang in the choir with our then girlfriend, later first wife. Looking through our translation of their research report, a longitudinal study of the effects of corporal punishment on children, they said (in Finnish, something like) "You might want to try translating the meanings of whole sentences, not individual words." Age-old advice, which in Chapter 2 we saw the DHP giving (Hofstadter 2009). We received it with some discomfort. This was a very friendly person who was giving us good advice; their body language was supportive, even maternal. But they were also suggesting that we had not done a good job translating their research report. This was professional guidance, instruction, from a kindly client; but it was also affect-regulation replicating the mother–child dyad. Our client was using body language as well as verbal language to teach us not only how to translate, but how to feel about translating—how to feel not only about the segmenting of the source text but about the SAP and the TRP. We didn't learn everything we needed to know from that encounter, but we remember it clearly as a major turning point in our apprenticeship as a translator, because that client's friendly body language helped us to transform our uneasiness at the implicit criticism into a determination to do better next time.

In Chapter 3 the TAP don't directly construct their own translator-function, of course—that is for others to do—but they do gradually, iteratively, strange-loopingly internalize various "standard" or "template" translator-functions that have been constructed by the culture, and position themselves vis-à-vis those normativized templates, cleaving to them or sheering off from them, thinking of them as selfobjects through and against which to define themselves. That self-definition process is not "punctual," of course, in the sense of happening once, at a specific point in time, or at a series of points in time—a series of jobs, say—but rather iterative in the strange-loops sense. One goes around and around with those selfobjects, seeing oneself taking up positions relative to them, feeling oneself coming into proximity and fending them off.

We haven't talked about translation norms in this book, but as Daniel Simeoni (1998) trenchantly showed some time ago, norms too have to be constructed, and constantly reconstructed, in very much the same iterative strange-loops manner as the translator-function (see Robinson 2003: 82–91 and 2020c, but also the Theo Hermans Pandemonium in section 2.4)—which is in any case a kind of norm-golem.

Presumably, too, the KMP's indignant anger at "critical theorists"/"constructivists" among translator scholars is triggered by a perception of their arrogant deviations from a specific norm-golem—one obviously related to thankless work maintaining social institutions, if our reading of the passages above was at all accurate, but also, perhaps, to empirical scholarship. Sentence 1 on page 144, y'all will recall, was the KMP's suggestion that "translation scholars rather work empirically to trace and account for the interfaces of translational and developmental actions"—not people, and not even actions, but interfaces between actions. The favored actions for both translators and development officials would apparently be self-abnegating actions in support of social structures. A big part of that self-abnegation would apparently be the "moratorium" on "judgment and criticism" mentioned in sentence 2—"until we know and understand enough." The important thing for translators, developers, and scholars is to suppress value-laden affect—judgment and criticism—and *work*, selflessly, for the good of all.

Our civic duty. And what's not to like about civic duty? It helps maintain civil society.

But perhaps there's more to it than that?

Conclusion: The Strange Loops of Translation as Transgressive Circulations: Johannes Göransson

Following the ideal model of strange loops launched by the DHP, the four chapters of this book have largely stuck to the range of translation strategies and conceptions considered normative: sense-for-sense and word-for-word translation in Chapter 1, the standard translator-function in Chapter 2, the periperformative management of translator norms in Chapter 3, and the embodiment of translation in Chapter 4. But does that cover the field?

Maybe not. An undercurrent of non-normative and anti-normative conceptions of translation has also been running all through the book: the half-suppressed emergence of a counterhegemonic pro-lesbian translator-function in Chapter 2; translators' periperformative resistance to performative regulation in Chapter 3; and the current of social affect that, surging through Chapter 4, channeled that periperformative resistance in Chapter 3 into the "locked-in" construction of reality.

By way of both looping the fraying strings of all four chapters back around on top of themselves and helping to fray them further, then, we propose to conclude by engaging the Johannes Göransson Pandemonium's (JGP's) *Transgressive Circulation* (2018) in dialogue. What's strangely loopy about the JGP's thin book is that it so gleefully shreds normativity, at once rejecting and reveling in all the abuse hurled at translators by its most strait-laced detractors: translators can't possibly reproduce a poem in another language; can't possibly get inside the mind of the source author; translation is a hoax, it's pure kitsch ... And Göransson says yes, and that's *how* it reproduces a poem in another language; that's *how* it infests the mind of the source author ...

C.1 Hoaxes

The DHP do not call strange loops hoaxes. They do call M.C. Escher's "Drawing Hands" a pleasurable hoax (2007: 103)—a hoaxical or illusionistic strange-loop imitation that only pretends to place each hand hierarchically above the other—but they then distinguish that hoax from "real" strange loops like Kurt Gödel's theorem. It's fortunate, they add, that non-illusory strange loops exist, because "the thesis of this book is that we ourselves—not our bodies, but our *selves*—are strange loops, and so if all strange loops were illusions, then we would all be illusions, and that would be a great shame" (103–4).

The DHP do, however, call the strangely loopy "I" "a tremendously effective illusion" (291–2). On page 104 it would be a shame if we were all illusions; on page 292 we are illusions, and it is sort of a shame, but what recourse do we have? On the next page the "I" is not only a hallucination but "a hallucination *hallucinated* by a hallucination" (293).

Actually, to be fair, the "I" that calls the "I" an illusion and a hallucination is not the authorial persona "Douglas Hofstadter" but one of the interlocutors in a DHP-indited dialogue. Only one DHP demon? That demon does in any case seize at least temporary control over the aggregate authorial DHP "I," because just a few pages later we find a non-dialogical DHP saying that "Ultimately, the 'I' is a hallucination, and yet, paradoxically, it is the most precious thing we own" (315), and calling their old friend the Daniel Dennett Pandemonium (DDP) to the stand to testify that the strangely loopy "I" is "a little like a bill of paper money," which feels real but in the end is just "a kind of illusion that we all tacitly agree on without ever having been asked" (315).

Presumably the difference between a hoax and a hallucination is that the former is a *deliberate* illusion, an attempt to fool people into believing in its reality, and the latter is an inadvertent illusion that can't really be avoided, unfortunate as that may be. So a hoax is an actionable deception and a hallucination is a mental disturbance that we can't control. To the extent that we take the illusion and the hallucination for reality, however, we're hoaxing ourselves, right? Or rather, our brain is hoaxing us: "In that sense," the DHP warn, "your brain has tricked itself" (291). What kind of trick? A practical joke? No: a magic trick. A "sleight of mind" (232). Not really a hallucination so much as a prestidigitator's illusion. A trick, a hoax. We mentioned in the Introduction (p. 6) that the DHP don't call the "I" an audience-effect, and speculated that they aren't really given to theatrical metaphors; but a magic show is also a theatrical event, and a magician's sleight-of-hand (or sleight-of-mind) illusion is also an audience-effect.

In fact, so far from celebrating the theatricality of such illusions, the DHP—whose Ph.D., let us not forget, is in physics—takes the only actual "reality" of human being to be the physical substrate of the brain (2007: 297). "I admit," they say, "that my stance has a definite ironic tinge to it. Sometimes the strict scientific viewpoint *is* hopelessly useless, even if it's correct. That's a dilemma. As I said, the human condition is, by its very nature, one of believing in a myth. And we're permanently trapped in that condition, which makes life rather interesting" (295). The only reality to human existence is the particle level: the DHP demons are cheerfully reductivist. What we believe about ourselves is our brain playing tricks on us, hoaxing us.

This to our mind is quite problematic, because while other DHP demons do speak of the "I" as "an *emergent* entity that exerts an upside-down causality on the world" (205; emphasis added), and call consciousness "an inevitable *emergent* consequence of the fact that the system has a sufficiently sophisticated repertoire of categories" (325; emphasis added), the DHP authoriality repeatedly gravitate away from an emergentist conception of the "I" as a new kind of volatile (or queered) reality emerging unpredictably *out* of the physical substrate of the brain, toward an unabashed reductionism that binarizes the simple reality of physics and the simple fakery of the "I" (see their celebration of Derek Parfit's reductionism, 310).

As we'll see, however, reading the DHP as limelighting the fakery of the "I" also resonates powerfully with the JGP on poetic translation/kitsch as a salutary hoax.

Not really a hoax, actually: an alleged hoax. A hoaxical accusation of hoaxing. A strange loop of hoaxicality:

> In modern and contemporary US literary discussions, translation itself is treated with a kind of dubiousness that suggests fear of a hoax. Translators who work on foreign writers that are not already very canonical are constantly faced with questions about the authenticity of their projects. How do I know that this foreign author is good enough to be translated? How can I tell if your translation is faithful? How do I know that the poet isn't just someone you invented? When I first began to translate Aase Berg's work, several people asked me if I had not simply made her up. To be a translator is to assume [the] role of a hoaxer, someone who might be undermining the quality and trustworthiness of literature (and taste).
>
> (Göransson 2018: 42)

Because "no one"—no "real person," real people obviously being defined as monolingual in English—can read the Aase Berg Pandemonium in Swedish, it is impossible to verify the quality of their work, or even their very existence. This normative imaginary posits the source author as the "true source" of the sought-after ontology behind the "I," but makes that source inaccessible: rather than the desired (imagined/normativized) direct access to the source, we have to settle for the strange loops of translation, which give us first-hand reports from an "I" who claims to be the author but speaks a language that is not the author's own, rendering the whole operation profoundly suspect. What if that source-"I" is a hoaxical bootstrap effort reverse-engineered by the (pseudo)translator?

But never mind. It's easy enough to duck out of such doubts: just read *great poetry originally written in English*. Ignore the parallel worlds of translated poetry and kitsch.

As the JGP reframe that normative imaginary, poetic translation and poetic kitsch have been "demonized," and mobilized as parallel and overlapping cautionary tales, for the purpose of protecting the supposed authenticity of "good" or "true" poetry. This would actually be the strange loop of demonization-cum-purification, which seeks to impose a good/bad or up/down or pure/base binary on a mixed field, with the aim of naturalizing the purified ideal as universal, stable, an ὄντως/*ontōs* that serenely and even smugly preexists every pitch that has ever pimped it. The effort to idealize/stabilize/universalize a putative source of pure ontological selfhood would have to proceed by denying and discrediting the strange loops by which it operates—precisely because those loops are the *only possible* way it can operate. Needless to say, as in Judith Butler's (1991) notion of panicked heterosexuality—and, by extension, the panicky clinging to any kind of normativity—that fatal reliance on strange loops of demonization and purification condemns the effort to failure; but the inevitability of that failure only goads it on to ever more frantic denials and denunciations, ever more intensified rhetorical attempts to naturalize its chosen fiction as the true originary source of all that is pure and wonderful. The accusation that translation is a hoax designed to corrupt the pure godlike authorial I-am-that-I-am is thus precisely the same kind of strangely loopy bootstrap operation as the supposed hoaxical corruption itself. As the JGP demons put it, "it may be useful to think about translation as a kind of sinthome for modern poetry: the counterfeit that must be posited in order for the authentic poem to exist" (45). ("Using Lacan, [Timothy] Morton [2007: 67] defines the 'sinthome' as 'the materially embodied, meaningless, and inconsistent kernel of 'idiotic enjoyment' that sustains an otherwise discursive ideological field'" [Göransson 2018: 45].)

The "materially embodied" abject: the JGP's champion. In an example they take from Daniel Tiffany (2014: 47–52), when William Wordsworth attacks the "gaudy and inane phraseology" of the gothic Graveyard poets, part of the accusation is that those poets import much of their kitschy sense of poetry from older and more ornate translations of Latin and Greek poems. Over here we see the William Wordsworth Pandemonium (WWoP) themselves as "'a man [just one!] speaking to men' in 'the language [just one!] really used by men' [not women]" (quoted in Göransson 2018: 40); over there, we find overblown poetic fancy talk borrowed from and thus corrupted by foreign poetic traditions. Good poetry is how good English "men" talk; bad poetry is how foreigners and women doll up ordinary spoken language with all manner of useless junk that is further corrupted by bad translators. (But, of course, in normative terms "bad translator" is a pleonasm. Translators are bad by definition.) Good poetry is pure; bad poetry/translation is corrupt. Good poetry conveys the pure interiority of one ordinary person to the pure interiority of another ordinary person; bad poetry is dragged through the mud, shit, and slime of debased/translated culture.

C.2 Interiority and Identity

But let's think more carefully about that "interiority" criterion for a purified identity. This is an idea(l) against which that the JGP demons rail over and over, and is also hinted at in the DHP's conception of the strange loops as undermining what they call the Capitalized Essence of Consciousness (Hofstadter 2007: 327–9)—the notion that each human being is born with a personalized dollop of consciousness which is *theirs alone*, and which therefore establishes them as a unique dualistic being, part body, part mind, part matter, part spirit. As long as the DHP demons go negative, their critiques line up nicely with avant-garde thought, in JGP and others; as we've seen, the "positive" exploration of the theatricality (performativity, phenomenology) of the "I" thus loopily constituted remains entirely alien to the DHP.

But then the avant-garde has never felt entirely comfortable with such positivities either. Avant-garde theatricality has historically been radically disruptive of identity and meaning. Indeed, the JGP demons theorize what John Durham Peters (2001: 38) calls transgressive circulations as undermining mental meaning-based interiority, or "the illusion of monoglossic completion and isolation" (Göransson 2018: 12). Loops and circulations, strange and transgressive, versus the Single Stable Self. This far, it would seem, the DHP would go as well. Significantly, however, the JGP also seek to undermine "the communication of interiorities, or as Peters puts it, the 'accurate sharing of consciousness'" (12; quoted in Göransson 2018: 20), and thus, apparently, to undermine the DHP's theory, aired here in section 1.3, that they and their wife Carol shared a consciousness that survived Carol's death, which we suggested might be mobilized with some revisions for sense-for-sense translation.

But the undermining effect is only apparent. In their project the JGP draw heavily on the work of the Daniel Tiffany Pandemonium (DTP), especially in *Radio Corpse* (1998) and *My Silver Planet* (2014), in particular strangely loopy notions like the suggestion that "the 'accursed share' produces pleasure, though it is founded on a logic of prohibition by swallowing the other, by contaminating itself and traversing its own boundaries" (1998: 161, quoted in Göransson 2018: 12). Because swallowing the other makes the other part of the self, the "I" is then also swallowing the prohibited self and contaminating it with the abjected outside, with the excluded and banished abject. This cannibalistic ingestion/introjection traverses boundaries from the outside in, but Haroldo de Campos's notion of baroque poetry/translation as cannibalism then immediately loops the ouroboric devouring of boundaries around into an outward movement as well, contaminating the abjecting force—say, literary right-mindedness—with pus and projectile vomit.

Despite appearances, in other words, it does seem that the JGP's "logic of prohibition" might well align with something like the DHP's strangely loopy account of partial marital fusion, with an avant-garde twist: following the JGP's embrace of the DTP, we would need to imagine the Carol demons not only as *ingested* by the DHP demons but as *prohibited by* and *contaminating* the DHP demons (and, for that matter, vice versa).

Imagistically, in other words, the metaphorical space populated by the JGP's avant-garde conception is a very different world from the one depicted by the DHP, where Kurt Gödel trashes the monolithic project of Russell and Whitehead in *Principia Mathematica*, but does so through the hygienic channel of subversive but equally elegant math, and where Johann Sebastian Bach creates strange loops in a deeply spiritual Lutheran apotheosis of baroque music. The DHP demons are rebels, iconoclasts; they take great pleasure in undermining and overturning the traditional pieties about consciousness, and in that they uncannily approximate the avant-garde critiques. But they tend to launch their incendiary campaigns from within a high-cultural enclave that is everywhere marked as explicitly and comfortably middle-class. A good fifth of *I Am a Strange Loop* (2007: 207–74) is a powerfully moving paean to the mutually enmeshed love the DHP and their wife Carol felt for each other before (and even after) Carol's tragically early death in 1993, by way of demonstrating their point that strange "I"-loops tend to originate in individual bodies but are not confined to the bodies in which they originate—that we can share "I"-loops, like the selves and nonselves that traverse inside-outside boundaries in the JGP. Unlike the JGP's transgressive circulations, however, there are no disgusting abject fluids in the DHP's strange loops. There is no rotten fruit. There is death, but it is ennobled; there is no carnivalistic reveling in it.

The JGP's focus is lower down:

> What might translation have to do with this "carnivalized," boundary-devouring, postcolonial baroque? Part of the answer can be found in Frost's "lost in translation" definition of poetry and in the oft-repeated truism that translation is "impossible." In his essay "Impossible Effigies," Daniel Tiffany [1998: 187] notes that
>
>> the impossible, as Georges Bataille observes, overwhelms utility, truth, and meaning, and thus engenders through translation unspeakable (and inconceivable) forms of exchange. Indeed, the impossibility lodged within translation is itself death, madness—and originality.
>
> Rather than the perfected "well-wrought urn," translation becomes an effigy that puts in motion a strange economy; instead of a "restricted economy," it generates a "general economy." In this strange economy, we may not only interact with the alien world by absorbing or rejecting, appropriating or conquering, but instead by becoming alien to ourselves—and opening up new realms of sensory-overwhelming "[orgies of] verbomania" [I.A. Richards' (1956: 11) derogatory term for bad writing]. Instead of going through the text to find a communication ideal in its interiority, we have to devour the text carnivalesquely.
>
> (53)

This "strange economy" of which the DTP and the JGP write would be a viral aesthetic/cultural proliferation of the DHP's strange loops. If as the JGP suggest anti-kitsch rhetoric is launched as "a defense against the foreign" but circulates eventually into "an immersive aesthetics, an aesthetic that revels in the weird, the strange, the night, the poetic, the possibly inhuman element of art," then kitsch *must* be abjected, "because it is a site where boundaries are traversed: the 'foreign body' of kitsch is 'lodged' inside the system of tasteful, modernist art. It may be foreign but it is inside the system, traversing the boundaries between inside and outside, creating folds in the system" (55). The "foreign body" that the DHP find inside the strange loops of their own

"I" is the "I" of their dearly departed wife Carol, who thus sadly/wonderfully survives her own death; the "foreign body" that the JGP (2018: 39, quoting Tiffany 2014: 1) find inside "the system of tasteful, modernist art" is kitsch, which is "'at once parasitic, mechanical, and pornographic,' associated with homosexuality and moral decay," and ultimately with bodily decay as well, disease and death.

Tellingly, too, the parasitism of kitsch is itself a strange loop in which, according to the Timothy Morton Pandemonium (TMP), "the aesthetic itself is … just a *disavowal* of kitsch that is, uncannily, its inner essence" (2007: 154; emphasis added). As the JGP comment: "Any work of art—no matter how lofty—becomes kitsch if you are immersed in its idiotic pull: 'All art becomes (someone else's) kitsch'" (2018: 45, quoting Morton 2007: 154). The attempt to split kitsch off from art is another strange purification/demonization loop. Every time one thinks one has banished kitsch from art, one loops around and finds the kitsch one banished inside the art one is trying to purify. By the JGP's/TMP's/DHP's queer logic, kitsch is loopily constitutive for art.

The JGP's main theme is the transgressive circulation of kitschy bad taste through various literary systems by translators and poets who themselves create partly by translating. What is circulated transgressively does not, however, corrupt and pollute literary systems alone:

> But "bad taste" can have a radical, political dimension. As recent discussions in biology have shown, there are more bacterial cells than human cells in the human body—more "foreign bodies" inside our human bodies than human bodies. These findings have led Timothy Morton to postulate about "nonselves" that are enmeshed in our surroundings rather than separate from it. Joyelle McSweeney [2014] has posited a "necropastoral" of poetics, "a political-aesthetic zone in which the fact of mankind's depredations cannot be separated from an experience of "nature" which is poisoned, mutated, aberrant, spectacular, full of ill effects and affects. This necropastoral zone "does not subscribe to humanism but is interested in non-human modalities, like those of bugs, viruses, weeds and mold." The necropastoral devours boundaries—between self and not-self, dead and alive, nature and human—and creates something like the baroque of De Campos [on poetry/translation as cannibalistic]. We might say that translation is like recent ecological thinking in its troubling of humanist boundaries.
> (55)

What is circulated transgressively also pollutes speciation. What the TMP and Joyelle McSweeney Pandemonium (JMP) help the JGP imagine is a strange loop of the entire planet, by which everything becomes everybody and everybody becomes everything, dead and alive. If the "I" of every human is "a hallucination *hallucinated* by a hallucination" (Hofstadter 2007: 293), which is to say a hoax or a simulacrum shammed up through a proliferation of strange loops, Gaia is the myth of a planetary goddess-"I" shammed up through strange loops in which everything that has ever existed participates.

C.3 Transminoritization

One theme the JGP invoke for this image of the kitschy poet/translator is the Deleuze&Guattari Pandemonium's (D&GP's) "minor" or minoritarian writer (54, 84):

> Chacun doit trouver la langue mineure, dialecte ou plutôt idiolecte, à partir de laquelle il rendra mineure sa propre langue majeure. Telle est la force des auteurs qu'on appelle « mineurs », et qui sont les plus grands, les seuls grands: avoir à conquérir leur propre langue, c'est-à-dire arriver à cette sobriété dans l'usage de la langue majeure, pour la mettre en état de variation continue (le contraire d'un régionalisme). C'est la langue majeure pour y tracer des langues mineures encore inconnues. Se servir de la langue mineure pour *faire filer* la langue majeure. L'auteur mineur est l'étranger dans sa propre langue. S'il est bâtard, s'il se vit comme bâtard, ce n'est pas par mixité ou mélange de langues, mais plutôt par soustraction et variation de la sienne, à force d'y tendre des tenseurs.
> (Deleuze and Guattari 1980: 133; emphasis in original)

> One must find the minor language, the dialect or rather idiolect, on the basis of which one can make one's own major language minor. That is the strength of authors termed "minor," who are in fact the greatest, the only greats: having to conquer one's own language, in other words, to attain that sobriety in the use of a major language, in order to place it in a state of continuous variation (the opposite of regionalism). It is in one's own language that one is bilingual or multilingual. Conquer the major language in order to delineate in it as yet unknown minor languages. Use the minor language to *send the major language racing*. Minor authors are foreigners in their own tongue. If they are bastards, if they experience themselves as bastards, it is due not to a mixing or intermingling of languages but rather to a subtraction and variation of their own language achieved by stretching tensors through it. (Massumi 1987: 116; emphasis in original)

In the terms we have developed out of D&GP minoritarianism (Robinson 2017b), the "minor" writer is not so much a stable identity as it is an audience-effect propelled by a series of loopy actions, so that the writer both *minoritizes* and *is minoritized by* readers. The vertiginous strange loops of minority generate the ostensibly stable "I" of "the minor writer" or minoritarian identity, but the strange loops by which the writer minoritizes and is minoritized are relentlessly self-defeating, and in that lies their transformative power.

By the same token the self-defeating strange loops by which the translator minoritizes and is minoritized are *transminoritizations*. When the Don Mee Choi Pandemonium (DMCP) call the Kim Hyesoon Pandemonium's (KHP's) poetry "a poetry of translation" (quoted in Göransson 2018: 82), they are attributing to the KHP too a transminoritizing impulse. As we hinted in the Introduction (pp. 19–21)—the hint strapped there to the bellies of "archaizing" sheep—the DRP demons found the same transminoritizing impulse in the Aleksis Kivi Pandemonium (AKP), who created Finnish literature out of their readings of European classics (Shakespeare, Cervantes, Dante, Homer) in Swedish translation, but also the Ludvig Holberg Pandemonium (LHP) in Danish (see Robinson 2017b: 87–8). The AKP wrote their first great play in the JGP's Swedish, then the colonial power language, but on the advice of the Emerson of Finnish nationalism, the J.V. Snellman Pandemonium (JVSP, 1806–81), rewrote it in subaltern (but emergent) Finnish as *Nummisuutarit* (1864, *Heath Cobblers* in our 1993a translation).

The normative translators in Chapter 2 who under the rubric of "the translator-function" find themselves compelled to ignore the strange loops of self-referential, incoherent, and archaized texts would be *transmajoritizers*. Because their task is to serve as the humble handmaidens of source-cultural greatness, and to that end to reproduce the brilliant interiority of the source author as faithfully as possible—a reproduction that, alas, is *by definition* inferior to their originals—theirs is not to question why.

Transminoritizing demons by contrast participate (see Robinson 2021b). They are thinkers and doers, and they insert themselves into the strangely loopy analogies between thinking and saying-as-doing by proliferating those analogies strategically. If they find their Source Author Pandemonium minoritizing in messy, out-of-control ways, they rejoice, and mess things up even more radically. Rather than shying away from the problems of self-reference, for example, they tend to exaggerate them, highlight them—by using, say, the c-model (Introduction, p. 11), where the Martin Luther

Pandemonium's English-speaking demons impossibly describe their attempts to translate the Bible into English (see Robinson 1997b: 95–6). Rather than tacitly fixing up the incoherent source text, if they suspect that the author is deliberately minoritizing—*doing* something with "bad writing" or "bad taste"—the transminoritizing demons jump in with both feet, go the writer one better.

C.4 Salutary Failures

Writing of the DMCP's brilliantly disturbing translation of the KHP's poetry from the Korean in *Mommy Must Be a Fountain of Feathers* (Choi 2008), which they edited and published in Action Books, the JGP note that "Choi's translation is not the attempt to 'recreate' an 'original,' but to bring it—its blackened, obscene vision, its 'sheer stuff,' its kitsch, its 'filthy things,' its 'fat,' its wounds, its grotesque body and language—across the bridge built by colonial politics and global economics between Korean and English" (82–3). It is an attempt to *do* things to English-speaking target readers across that bridge—an attempt launched in some sense by both Korean women, though technically the KHP are the "source author" and the DMCP are the "translator":

> In this gothic vision of poetry and translation, Kim and Choi become uncanny doubles, "two daughters too many," an excess. Choi recounts that when Kim was asked how she felt about being translated, Kim replied, "it's like meeting someone like myself." The blackened realm of poetry generates doubles, too-much-ness. In the blackened space where poetry is translation and vice versa, Choi "dwells" in Kim's "house," that constantly ruptured space where the I tries and fails to imagine itself as a space, that space that cannot keep out history's political violence. And because poem and translation dwell in the same blackened space, the translation "exists on the same plane as the original poem," not as an inferior (or superior) version.
>
> (Göransson 2018: 83)

"That constantly ruptured space where the I tries and fails to imagine itself as a space" is a space of nightmarishly proliferating strange loops that both consolidate and rupture, both build and tear down. "That space ... cannot keep out history's political violence," not just because political violence is a viral proliferation of strange loops that build and destroy, but because the "I" that "tries and fails to imagine itself as a space" is complicit in that violence, is part of it, shares strange loops with it. Like the bacteria that outnumber us in our own bodies, like the "non-human modalities ... of bugs, viruses, weeds and mold" of which the JMP write, the sociopathologies of History Я Us. History, like nature, is "poisoned, mutated, aberrant, spectacular, full of ill effects and affects"—and those ill effects and affects shape us, because they are inside us, and we are in them. We are necropastors, blackening the future.

Implicit in transminoritizing is a very different conception of the translator:

> Unlike translation advocates such as Venuti, who insists on making translators legitimate,[1] Choi insists on the translator's lowliness, her "failure," her abjectness. In "Freely Frayed+=q + Race =Nation," she writes: "It turns out that I'm a mere imitator, the lowly kind, which is none other than a translator, a mimicker of mimetic words ... I twirl about frantically frequently farfar to the point of failure feigning englishenglish." Choi positions the translator as a failure—for example a failure to properly master the Korean with masterful English, leading to "englishenglish" instead of "English"—but it's exactly in this failure that the strength of her translations, as well as Kim's originals, comes from. Choi writes: "translation is a process of perpetual displacement." She writes:
>
>> The displaced poetic identity persists in this dislocation, translating itself out of the orders of darkness through the translator; another displaced identity. We have no choice but failfail. Failingfailing, it's painful becoming a translation, becoming an immigrant. [Hix and Choi 2015: 6[2]]
>
> Choi participates in Kim's transgressive circulation, a circulation that ruins the private space, the clean space of patriarchy. As she does, she suffers from exactly the kind of "foreign influence" Steiner fears: the foreign language corrupts and disables her, causing her to become a "failingfailing" translator—both invisible and hypervisible like the "blackened space" of Kim Hyesoon's poetry.
>
> (83–84)

In almost every sense this vision of translation-as/of-poetry as "a process of perpetual displacement" tracks with the DHP's disturbing vision of the creation-cum-displacement of the "I" through strange loops. In the end the JGP's "transgressive circulations" *are* strange loops. If the displacing and dislocating and corrupting and disabling of identity invisibilize the translator, it also renders them "hypervisible." That hypervisibility may be the visibility of a fire-blackened site where once stood a house, drawing the lurid gazes of rubbernecking passers-by; but a displaced and dislocated identity is still an identity of sorts, a house of cards that *feels* real enough, until the next fire sweeps through, or a wisp of a breeze sneaks in through an open window. The DHP are obsessively concerned with "reality" and imposture, the ways in which the personal identity limned in by and through strange loops pretends to be real, though it is just an illusion; to that epistemological concern the DMCP, and through the DMCP their editor the JGP as well, add a nightmarish imagery of failure and pain and corruption and disability, the "I" as translation, the "I" as immigrant, the "I" as refugee (Robinson 2020a), all summoned forth, like hard-won victories, from the orders of darkness. The translational immigrant identity painfully built up through transgressive circulations is not just fakery; it's failed fakery.

In the DHP, by contrast, the only real problem seems to be that the "I's" fakery is *too* successful. It seems so real that it's difficult for them to convince the skeptics of its falsity. There's really nothing else wrong with it. Certainly, there is no abjection to cope with. There is pain, perhaps—the pain of loss—but that doesn't seem to undermine the DHP's identity the way the KHP's and the DMCP's pain as Korean women undermines theirs. One is tempted to say "privilege"—the privilege of a white middle-class American male intellectual who has achieved considerable success and even fame as a writer and a thinker—but that seems too glib. The DHP are more pandemonially complex than that. There is, after all, something in their work that resonates powerfully with the JGP's transgressive circulations.

Maybe what the DHP demons are faking is the failure to fail?

Notes

Introduction

1   The DHP (Hofstadter 2007) do, however, explicitly mobilize the cultural phenomenon of literary translation, exactly twice:

   No one has trouble with the idea that "the same novel" can exist in two different languages, in two different cultures. But what is a novel? A novel is not a specific sequence of words, because if it were, it could only be written in *one* language, in *one* culture. No, a novel is a *pattern*—a particular collection of characters, events, moods, tones, jokes, allusions, and much more. And so a novel is an abstraction, and thus "the very same novel" can exist in different languages, different cultures, even cultures thriving hundreds of years apart. (224)

   If the molecules making you up are *not* the "enjoyers" of your feelings, then what is? All that is left is *patterns*. And patterns can be copied from one medium to another, even between radically different media. Such an act is called "transplantation" or, for short, "translation."

   A novel can withstand transplanting even though readers in the "guest language" haven't lived on the soil where the original language is spoken; the key point is, they have experienced essentially the same phenomena on their own soil. Indeed, all novels, whether translated or not, depend on this kind of transplantability, because no two human beings, even if they speak the same language, ever grow up on exactly the same soil. How else could we contemporary Americans relate to a Jane Austen novel? (257)

   The abstract reductivism of the notion that "a novel" and "feelings" are merely cognitive "patterns" will receive close scrutiny in Chapter 4. The important point to note here is that there is no explicit strange loopiness in these two analogues.

2   The Aristotle Pandemonium don't exactly spell out this speaker-audience *ēthos*-collaboration in the *Rhetoric*, but their argument can be read that way; see Robinson (2016b: 9, 59, 146, 190) for discussion.

3   Or "in traducing me you inevitably traduce me": in English to traduce is both "to translate/transmit/pass on" and "to malign/slander/betray." The problem with using that homonymy to translate the Italian source text, of course, is that it doesn't specify the accusatory directionality: "traducer, traducer" could mean "to betray is to translate" just as well as it might "to translate is to betray."

4   For a more recent case, see the DRP's transcreation of Volter Kilpi's unfinished posthumous final novel, *Gulliverin matka Fantomimian mantereelle*, as *Gulliver's Voyage to Phantomimia* (Robinson 2020d). Because the Volter Kilpi Pandemonium (VKP) invoked Swift's found-manuscript meme and claimed that Gulliver's bound manuscript appeared on their desk in the university library, *in English*, and they themselves translated it into Finnish, the DRP realized that they could pretend to have found the same manuscript and *edited* it in English—though they would actually be translating the part that the VKP finished and finishing the part the VKP left unwritten at their death in 1939. In addition to the *Pale Fire* fun this led to, the DRP were able to build into the transcreation many temporal loops that proliferated the VKP's initial epistemological play many-fold. See Robinson (2022b: Finale) for discussion.

5   One of the notable battles fought over this issue of flattening versus heightening translations has concerned the Aleksandr Sergeyevich Pushkin Pandemonium's (ASPP's) *Евгений Онегин/Yevgeny Onegin*—mainly because the Vladimir Nabokov Pandemonium (VNP) famously attacked the heighteners (1941, 1955) for deviating from Pushkin's text and offered their own aggressively flattened translation (1964) as a corrective. The sticking point for translations of the ASPP's works is always the extreme difficulty of rendering their easy and casual and apparently colloquial but also brilliantly heightened verse with the same level of casual ease *and* apt brilliance—while also following the verse form. To follow the verse form, the established view goes, one has to deviate from the meaning of the verses—and even then it's almost impossible to capture the ASPP's wonderfully blithe insouciance. As it happens, the DHP tried their hand at an *Onegin* translation (Hofstadter 1999) around the time *Le Ton Beau de Marot* (Hofstadter 1997) was coming out, and worked very hard there to maintain the verse form—but also signally flattened the ASPP's style. In their Translator's Preface (Hofstadter 1999: xxiii–xxvi) they reviewed the VNP's infamous 1964 *Onegin* at some length, clearly appalled at the VNP's manifest effort to suck all pleasure out of the English text, in order to show that the ASPP's novel must be read in Russian (and many critics have agreed; see, e.g., Rosengrant 1994; Dolinin 1995; Trubikhina 2008; and Razumnaya 2012). But frankly, and disappointingly, the DHP's *Onegin* struck us as more similar to the VNP's than different. Yes, the VNP's much-vaunted "literalism" was not so much word-for-word translation as a refusal to employ any of the ASPP's poetic heightening strategies (rhyme, meter, playful mixtures of archaic, poetic, and colloquial diction); the DHP, by contrast, flattens the text in diction only, reproducing the *Onegin* stanza form accurately and perhaps seeking to imitate the ASPP's poetic flair, but lacking the poetic flair to pull it off. Unfortunately, the flat mundanity of both translators' diction tends to override any poetic heightening the rhyme and meter might otherwise have imparted. For example, the first three lines of Chapter 4, sonnet 18:

ASPP

Вы согласитесь, мой читатель,
Что очень мило поступил
С печальной Таней наш приятель …

VNP

You will agree, my reader,
That very nicely did our pal
Act toward melancholy Tanya …

DHP

Dear reader, would you be conceding
That with poor Tanya our Eugene
Behaved himself with style and breeding? (57)

The VNP and DHP there both resound with a comparable flatness, despite the VNP's self-proclaimed aim of reducing the novel to the ugliest possible literal rendition and the DHP's horror at literalism. In fact, the VNP's translation is "literal" only in a very loose sense; this would be closer:

VNP[DRP]

You will agree, my reader,
That very kindly behaved
With sad Tanya our homie

Comparing the VNP with the DHP: in Russian Tanya is sad, or doleful, or the VNP's "melancholy," not the DHP's "poor"; and while the VNP's "our pal" is perhaps a bit strong for "наш приятель," the DHP's "our Eugene" is a bit weak. The DHP differs from the VNP mainly, however, in the distortions required to get the rhyme: "be conceding" and "with style and breeding." "To concede" works as a verb for "согласиться/*soglasit'sya*"/"to agree," but the continuous tense is awkward, and "with style and breeding" is a questionable expansion of "мило/*milo*," which is "nicely" or "kindly." The fact is, the Eugene Onegin Pandemonium (EOP) never behave "with style and breeding." They are almost always blunt and rude. They don't know how to behave. The best they can manage here is *not* to be curt with Tanya—so, nicely. Kindly. Not "with style and breeding." But the DHP need that expansion for the rhyme. And yet they write:

> One of the most central maxims that I've tried to abide by is that of *covering my tracks* as far as rhymes are concerned. By this, I mean that a reader should not be able to tell, by looking at two rhyming lines, which came first and which was created later, for the sake of rhyme. Both lines should seem equally natural; neither should suggest that it was written just to rhyme with the other—even

though it might very well have been. One does one's best to make each line look as though it, in and of itself, was the optimal way of packing the thought in words, and as though it came effortlessly. (1999: xxxv)

Never mind that "be conceding" and "with style and breeding" fail this test hopelessly; the main problem there is that the DHP are not so much writing poetry as "covering their tracks," which is to say, trying to hide the fact that they are not a poet. The English *Onegin* that the DHP love, the one by James Falen (1990) that they read out loud with their wife Carol long before conceiving the project of translating it themselves, doesn't rhyme by "covering tracks": it's poetry. Every rhyme, every accommodation to meter seems not only effortless but a glorious opportunity to celebrate the prosodic resources of the English language. This is not an invidious comparison with anything the DRP might have created; we are not poets either!

For example, a colloquial prose translation might go something like "You'll have to agree, dear reader, that our guy treated sad Tanya not half-badly." A closer or more "literal" rendition might go like this:

DRP[1]
You would acknowledge, gentle reader,
That very kindly has behaved
Toward teary Tanya our loss leader.

"Loss leader"? That's not literal; that's just for the rhyme. A case could be made for it—the ASPP portray the EOP as the *primus inter losers*, so to speak, and are a bit surprised here that the EOP weren't nastier with Tanya—not to mention that "loss leader," hard on the heels of "Toward teary Tanya," is alliterative—but we too are covering our tracks. It ain't poetry.

So, a little looser:

DRP[2]
But reader, don't you have to credit
Our sidekick for the kindly way
He handled Tanya's plea pathetic?

The alliterative inverted word order in "plea pathetic" is poetic heightening, to be sure, but perhaps not really justified here: again, it's for the (near-)rhyme. The DHP campaigns hard against near-rhymes (Hofstadter 1999: xxxvi–xxxvii) as "not rhymes at all," apparently not knowing the history of English poetry, which *prefers* near-rhymes to the doggerel taste of exact rhymes, or the history of Russian poetry, where exact rhymes are not only preferred but fairly easy to produce without syntactic contortions. But that still doesn't make DRP[2] poetry. We're still covering our tracks. One more:

> DRP³
> You'll acknowledge, target reader,
> That very nicely have the D-
> RP be-englished *E. Onegin.*

Come on, forget track-covering, that doesn't even rhyme! (It's a self-defeating/self-ironic/self-deprecatory strange loop.)

6   The DHP mention that seminar in the Preface to the *Onegin* translation:

> In the spring semester of 1997, just as *Le Ton Beau de Marot* was about to appear, I offered a seminar on verse translation at Indiana University, and among the many works we were looking at, *Eugene Onegin*—in these four highly diverse anglicizations—featured prominently. I felt the best way to compare the four approaches was to ask each student to concentrate on one stanza, studying it carefully in all four translations, making notes about prominent merits or defects, and then coming to class and performing the four rival stanzas out loud with as much skill as they could muster, after which a class-wide discussion would take place. (1999: xiii)

During our visit to that seminar there was no discussion of *Onegin*—but then we didn't start studying Russian in earnest until a year and a half later, when we met the Russian woman who became our wife.

Chapter 1

1   See section 2 of the "Hermeneutics" entry in Robinson (2021a) for a discussion of homophonic translation, especially David Melnick's *Men in Aïda*; for a discussion of radically literal translations of заум/*zaum* "beyonsense," see Robinson (2017e: 84–6).
2   For a recent linguistic reading of Jerome's "exception" along these text-type lines, see Munday (2001/2012: sec. 2.1).
3   We are also, here, in the vicinity of so-called "spirit-channeled" translations, explored in Robinson (2001): the notion of "sharing demons" with a dead source author offers a possible explanation of the feeling some translators have of channeling the dead spirit of the source author when they translate. The legend of the Holy Ghost Pandemonium's mystical translation of the Septuagint through the bodies of the Seventy translators supposedly gathered at Alexandria in or around 280 BCE—which the ESHP categorically rejected, but implicitly invoked in piously claiming that in Holy Scripture "even the syntax contains a mystery"—is one of the examples discussed in Robinson (2001: 49–54).
4   Hofstadter (2009: 97–8) does pause near the end of their book to consider translation as "co-creation"; but rather than exploring that possibility on the model

of the partial fusion of Doug's Pandemonium with Carol's Pandemonium, they invoke the metaphor of making a baby: the author is the mother, who does most of the creative work, but the translator is the father, "and though his procreative role is certainly far subsidiary to the mother's, it is nonetheless indispensable" (98). In intracytoplasmic sperm injection (ICSI), of course, all that is needed from the father is a single sperm cell—no sex needed—so when the DHP also launch the parallel metaphor of food, saying that the author-mother cooks the meal but the father-translator can still provide a little spice (98), they are not necessarily extolling the spicy properties of ejaculate. Doug and Carol made two babies, so perhaps Hofstadter is thinking more holistically here, and possibly even tacitly including the feeling they had of pandemonial fusion; but not engaging that feeling explicitly, and so relegating translational co-creation metaphorically to the mere machinery of sex, or even of IVF/ICSI, seems like a missed opportunity to us.

5   The DHP (Hofstadter 2007: 152) go on: "Sometimes the characters are completely unaware of Situation B, which can make for a humorous effect, whereas other times the characters in Situation A are simultaneously characters in Situation B, but aren't aware of (or aren't thinking about) the analogy linking the two situations they are in. The latter creates a great sense of irony, of course."

For discussions of the translationality of such analogical plot doublings in Shakespeare, see Gentzler (2017) on the Pyramus and Thisbe play and similar reduplications in *A Midsummer Night's Dream* (32–42) and "The Murder of Gonzago" in *Hamlet* (191–5).

Chapter 2

1   Like us that same year, the Arrojo Pandemonium chide the Díaz-Diocaretz Pandemonium for blurring their development of the "translator-function" out of the Foucault Pandemonium: "Unlike Myriam Díaz-Diocaretz, whose conception of 'translator-function' (1985: 24–41) is vaguely reminiscent of Foucault's, I explicitly relate the notion of 'translator-function' which I propose here to Foucault's seminal essay 'What's an Author?' ([Harari] 1979: 141–59)" (1997: 31n2).

2   The Charles Sanders Peirce Pandemonium (CSPP) floated the idea of "endless semiosis" in 1866, and didn't begin to see it as a red herring until 1903 (1931–58: 2.242, 2.275); later that year they declared their old theory absurd, but weren't able to work out a pragmatic solution to the problem immediately. As Short (2004: 219–26; 2007: 53–9) shows, they solved it between 1903 and 1907, first by theorizing the interpretability of a sign as its immediate interpretant (1931–58: 4.536, 539 [1906]), then by realizing that *habit* imposes a good-enough end on semiosis: "it is the habit itself, and not a concept of it, that is the ultimate interpretant of a concept"

(Short 2007: 58). But of course the pragmatic solution is not nearly as sexy as the original formulation, as it was picked up by the Jacques Lacan and Umberto Eco Pandemonia and turned into a kind of signature trope for poststructuralism.

Chapter 3

1. Translating *das Man* as "the 'they'" does get at something like the unconscious dissemination of self-structuring throughout a whole social field that the MHP mean by *das Man*; unfortunately, it also lexically redirects the depersonalized/generalized speaker-*inclusion* in "one" to a depersonalized/generalized speaker-*exclusion* in "they." *Das Man* sets up a reciprocal exchange between the personal first-person "I" and the impersonal third-person "one"; "the 'they'" shifts that all over to shadowy others who control things from behind the scenes—which is not what the MHP is theorizing at all.

    Similar problems beset other translations that have been offered, such as "the Everyone" (Funkenstein 2000: 178) and "the Anyone" (Blattner 2007). Both indefinite pronouns (see Haspelmath 1997) come closer to *das Man* than "the 'they'" in suggesting that the MHP's concept is a kind of collectivized self, though neither stands in intimate relation with the self; the focus in both is on collectivization rather than the operation of that process in individuals. The big problem with "the Everyone" and "the Anyone," however, it seems to us, is that neither is integrally connected with periperformatively *guided choice*, specifically crowd guidance to the selection from among ethical options of the normative one. *Das Man* may be nearly anonymous like "everyone" and "anyone," but it *wants* things, and puts pressure on individuals to want those things too. "The Everyone" is literally everyone, the whole world, all human beings, and therefore incapable of helping the individual choose a course of action. Indeed, the childhood cry "But everyone does it!" would seem to invoke something like the exact opposite of *das Man*—random, indiscriminate action sanctioned by an unregulated mass—as a channel of resistance to precisely the kind of periperformative (in this case parental) pressure to adhere to group norms that *das Man* channels. "The Anyone" makes every individual interchangeable, which is not so very different from the kind of conformity *das Man* seeks to create; but not *normatively* interchangeable, because we've all been ethically regulated by society, as *das Man* would prefer, but *randomly* interchangeable, simply because we're all human. Because groups proliferate and overlap, of course, there can be no universal set of norms to which *das Man* seeks to make us conform; and because the ethical pressure that is *das Man* works imperfectly on imperfect beings in an imperfect world, conformity is never total even within a single group.

2   For feminist reactions to the George Steiner Pandemonium on translation-as-rape, see, e.g., Bassnett 1992; Chamberlain 1992; and Simon 1995; for a feminist commentary on and complication of this reaction, see Arrojo 1995.
3   The tension here is between a linguistic approach to textual equivalence, for which syntactic and semantic comparison between the source and target texts is not only "enough" but *everything*, and a *skopos* or "action-oriented" approach to the professional networks that bring translations into existence (see e.g., Pym 2010: chs. 2–3 for linguistics, 4 for *skopos*).
4   While the Antoine Berman Pandemonium's (ABP's) title *L'Épreuve de l'étranger* was explicitly a French translation of Heidegger's/Hölderlin's "die Erfahrung des fremden"/"the experience of the foreign," I would be very surprised if somewhere in the subliminal underpinnings of the phrase in the ABP's mind was not the FSP's phrase "das Gefühl des fremden"/"the feeling of the foreign"—which does not appear in their book, but does seem nevertheless to inform both the FHP's/MHP's German phrase and the ABP's French title. After all, the ABP's book is among other things an updating of the FSP's 1813 Academy address (which they mysteriously date to 1823), and as we saw in Chapter 1 the National Translation Project that the FSP advocate there does feature "das Gefühl des fremden" as its primary periperformativity. Also, the ABP's brief section on the FSP in their tenth chapter (Heyvaert 1992: 141–52) reprises the FSP's pioneering work on hermeneutics, which—though the ABP doesn't mention this either—is *Gefühl*/feeling-based and borrowed from the *Gefühl*/feeling-based thoughts on hermeneutics from the true pioneer, the Johann Gottfried Herder Pandemonium. The fact that the ABP do not mention the feeling of the foreign even once suggests that their imagination, indeed like the FSP's own, was moving away from periperformativity; but the socioaffective convergence of the three phrases—the FSP's and the FHP's in German, the ABP's in French—is suggestive nevertheless.
5   Sakai (1997: 14) does come close to what we've called the "third-person we" in arguing that "in the case of translation, the oscillation or indeterminacy of the personality of the translator marks the instability of the we as the subject rather than the I." We would insist that the "we" is not in fact unstable; what is unstable in the "we" is the shift to "one."

Chapter 4

1   Interestingly, we learned recently that several of those photocopies ended up on the shelves of Finnish university libraries: in their Licentiate thesis at the University of Eastern Finland, studying the problem of translating between animal language and human language, Ella Vihelmaa (2018) cited that photocopy, with

a 1980 publication year. Intrigued by how they managed to find it, and where the "publication" year came from—the photocopy the DRP made had no such identifiers on it, just the title in Finnish and English and the author's name—the DRP posted on the find to social media, and sure enough, a week later Vihelmaa emailed to apologize for getting the year wrong, but 1980 was the guess some librarian had written on the copy they had on the shelf. Vihelmaa was able to report that *five* Finnish university libraries possessed copies of that photocopy!

2   This section is a radical rewriting of Robinson 2012b; for other reflections on the HMP on translation, see also Robinson 2014a and 2014b.
3   Icotic theory began to emerge as an extension of somatic theory in early drafts (from about 2009) of what eventually became Robinson (2016b); see also Robinson (2013b, 2016c, 2016d, 2017e, and 2019.)
4   In Finland the public postboxes are obviously and undeniably orange—except to Finns, who think they're yellow. They have a word for orange, *oranssi*, but it's clearly a foreign loanword, and they feel more comfortable *seeing* those postboxes as *keltaisia* "yellow." We once convinced a class of Finnish translation students to try the thought-experiment of experiencing postboxes as orange. They managed—but only after a good 30–45 minutes of working at it. That class was a complex relearning phenomenology.

Conclusion

1   The LVP (Venuti 1998) sought to shoehorn Deleuzean minoritarianism into their Schleiermacherian notion of foreignization by painting the foreignizing translator as a strong visible Romantic hero who boldly reproduces a minoritarian writer's deviances accurately. Unfortunately, as the DRP have shown at some length (Robinson 2011: ch. 5 and 2017b: 150–63), the differences between foreignizing and domesticating in the LVP are so imperceptible that both strategies wind up transmajoritizing the brilliant interiority of the stable source author. Robinson (2021b) also launches an avant-garde critique of the LVP's heroizing victim discourse about the poor invisibilized translator.
2   We assume that the print version of this book is a radically edited version of the digital version (or vice versa) that we found online, where that same passage reads like this: "My hope is that the displaced poetic or narrative identity manages to persist in its dislocation, translating itself out of the orders of darkness alone or with assistance from the translator who must also translate herself."

References

Andrew, Sally. 2019. *Death on the Limpopo*. Cape Town, SA: Umuzi/Penguin Random House.

Apter, Emily. 2013. *Against World Literature: On the Politics of Untranslatability*. London, UK: Verso.

Apter, Emily, Jacques Lezra, and Michael Wood, trans. 2014. Barbara Cassin, ed., *Dictionary of Untranslatables: A Philosophical Lexicon*. Translation of Cassin 2004. Princeton, NJ: Princeton University Press.

Arrojo, Rosemary. 1995. "Feminist, 'Orgasmic' Theories of Translation and their Contradictions." *TradTerm* 2: 67–75.

Arrojo, Rosemary. 1997. "The 'Death' of the Author and the Limits of the Translator's Invisibility." In Mary Snell-Hornby, Zuzana Jettmarová, and Klaus Kaindl, eds., *Translation as Intercultural Communication: Selected papers from the EST Congress, Prague 1995*, 21–32. Amsterdam, Netherlands, and Philadelphia, PA: John Benjamins.

Ashby, Hal, dir. 1979. *Being There*. Lorimar.

Austin, J.L. 1962/1975. *How To Do Things With Words*. Ed. J.O. Urmson and Marina Sbisà. 2nd edition. Cambridge, MA: Harvard University Press.

Aviram, Amittai. 2002. "The Meaning of Rhythm." In Massimo Verdicchio and Robert Burch, eds., *Between Philosophy and Poetry: Writing, Rhythm, History*, 161–70. New York, NY: Continuum.

Avtonomova, Nataliya. Unpub. "Filosofia, perevod, 'neperevodimost': kul'turnye i kontseptual'nye aspekty" ["Philosophy, Translation, 'Untranslatability': Cultural and Conceptual Aspects"]. Unpublished manuscript.

Baker, Mona. 2000. "Towards a Methodology for Investigating the Style of a Literary Translator." *Target* 12.2: 241–66.

Baker, Mona. 2006. *Translation and Conflict: A Narrative Account*. London, UK, and New York, NY: Routledge.

Bakhtin, Mikhail. 1929/2002. *Problemy poetiki Dostoevkogo* ("Problems of Dostoevsky's Poetics"). Moscow, Russia, and Augsburg, Germany: Werden. Online at http://ir.nmu.org.ua/bitstream/handle/123456789/35362/aabaf43b33b35a23cbedd506dfe7f04b.pdf?sequence=1. Accessed April 9, 2015.

Barthes, Roland. 1968. « La mort de l'auteur ». Source text of Howard 1967. *Mantéia* 5.

Bassnett, Susan. 1992. "Writing in No Man's Land: Questions of Gender and Translation." *Ilha do Desterro*, Universidade Federal de Santa Catarina #28 (2nd semester): 63–73.

Beebe, Beatrice, and Frank M. Lachmann. "Mother-Infant Mutual Influence and Precursors of Psychic Structure." In Arnold Goldberg, ed., *Progress in Self Psychology* 3: 3–25. Hillsdale, NJ: Analytic, 1988.

Benjamin, Walter. 1923/1972. "Die Aufgabe des Übersetzers." In Tillman Rexroth, ed., *Kleine Prosa, Baudelaire-Übertragungen*, 9–21. Vol. 4, Part 1 of Walter Benjamin, *Gesammelte Schriften* ("Collected Writings"). Frankfurt a. M., Germany: Suhrkamp.

Benveniste, Émile. 1966. *Problèmes de linguistique générale*, vol. 1. Paris, France: Gallimard.

Berman, Antoine. 1984. *L'Épreuve de l'étranger: Culture et traduction dans l'Allemagne romantique*. Paris, France: Gallimard.

Bethlehem, Louise Shabat. 1999. "Strange Loops and Writes-of-Passage: Double-crossing Diaspora." *South Atlantic Quarterly* 98.1/2 (Winter): 255–66.

Black, Scott. 2015. "Tristram Shandy's Strange Loops of Reading." *ELH* 82.3: 869–96.

Blattner, William. 2007. *Heidegger's Being and Time: A Reader's Guide*. New York, NY: Continuum.

Blum-Kulka, Shoshana, and Eddie A. Levenston. 1983. "Universals of Lexical Simplification." In Claus Faerch and Gabriele Kasper, eds., *Strategies in Interlanguage Communication*, 119–39. London, UK, and New York, NY: Longman.

Boisacq, Emile. 1916. *Dictionnaire étymologique de la langue grecque, étudiée dans ses rapports, avec les autres langues Indo-Européennes* ["Etymological Dictionary of the Greek Language, Studies in Relation to the Other Indo-European Languages"]. Paris, France: Librairie C. Klincksieck. Online at http://www.archive.org/stream/dictionnairety00bois/dictionnairety00bois_djvu.txt. Accessed September 16, 2019.

Borchardt, Rudolf, trans. 1923/1967. *Dantes Commedia Deutsch* ["Dante's *Comedy* in German"]. Vol. 14 of *Gesammelte Werke*. Stuttgart, Germany: Klett-Cotta.

Boulanger, Pier-Pascale, trans. 2011. Henri Meschonnic, *Ethics and Politics of Translating*. Translation of Meschonnic 2007. Amsterdam, Netherlands, and Philadelphia, PA: John Benjamins.

Bowker, Lynne. 2000. "A Corpus-Based Approach to Evaluating Student Translations." *The Translator* 6.2: 183–210.

Bowker, Lynne. 2001. "Towards a Methodology for a Corpus-Based Approach to Translation Evaluation." *Meta* 46.2 (June): 345–64.

Brennan, Eileen, trans. 2006. Paul Ricoeur, *On Translation*. Translation of Ricoeur 2004. London, UK, and New York, NY: Routledge.

Butler, Judith. 1991. "Imitation and Gender Insubordination." In Diana Fuss, ed., *Inside/Out: Lesbian Theories, Gay Theories*, 13–31. London, UK, and New York, NY: Routledge.

Cardozo, Mauricio Mendonça. 2019. "Translation, Humanities, and Contemporary Thinking: Towards an Understanding of Translation as Practice and Critique of Relational Reason." In Spitzer 2019a: 36–64.

Cassin, Barbara, ed. 2004. *Vocabulaire européen des philosophies (dictionnaire des intraduisibles)* ["European Vocabulary of Philosophies (Dictionary of Untranslatables)"]. Paris, France: Seuil.

Cassin, Barbara. 2016. "Translation as Paradigm for Human Sciences." *The Journal of Speculative Philosophy* 30.3: 242–6.

Chamberlain, Lori. 1992. "Gender and the Metaphorics of Translation." In Lawrence Venuti, ed., *Rethinking Translation: Discourse, Subjectivity, Ideology*, 57–74. London, UK, and New York, NY: Routledge.

Choi Don Mee, trans. 2008. Kim Hyesoon, *Mommy Must Be a Fountain of Feathers*. South Bend IN: Action Books.

Dai, Guangrong (戴光荣). 2019. 双语语料库在翻译质量评估中的研究 ["Using Bilingual Corpora in Evaluating Translation Quality"]. Second WITTA International Symposium on Translation Education, Kunming, Yunnan Province, PRC, July 4.

Damasio, Antonio R. 1994. *Descartes' Error: Emotion, Reason, and the Human Brain*. New York, NY: Putnam.

Damasio, Antonio R. 1999. *The Feeling of What Happens: Body and Emotion in the Making of Consciousness*. New York, NY: Harcourt.

Damasio, Antonio R. 2003. *Looking for Spinoza: Joy, Sorrow, and the Feeling Brain*. New York, NY: Harcourt.

Dawkins, Richard. 1976. *The Selfish Gene*. Oxford, UK, and New York, NY: Oxford University Press.

Deleuze, Gilles, and Félix Guattari. 1980. *Mille plateaux* ["A Thousand Plateaus"]. *Capitalisme et schizophrénie*, Tome 2. Paris, France: Minuit.

Dennett, Daniel. 1991. *Consciousness Explained*. New York, NY: Little, Brown.

Derrida, Jacques. 1967. *De la grammatologie* ["Of Grammatology"]. Paris, France: Minuit.

Derrida, Jacques. 1972. « Signature contexte événement » ["Signature Context Event"]. In Derrida, *Marges de la Philosophie*, 365–93. Paris, France: Minuit.

Derrida, Jacques. 1978. *Spurs: Nietzsche's Styles/Éperons: Les Styles de Nietzsche*. Translated by Barbara Harlow. Chicago, IL: University of Chicago Press.

Derrida, Jacques. 1996. *Le Monolinguisme de l'autre; ou, la prothèse d'origine* ["Monolingualism of the Other; or, the Prosthesis of Origin"]. Paris, France: Galilée.

Diamond, Marian C., David Krech, and Mark R. Rosenzweig. "The Effects of an Enriched Environment on the Histology of the Rat Cerebral Cortex." *Journal of Comparative Neurology* 123 (1963): 111–20.

Díaz-Diocaretz, Myriam. 1985. *Translating Poetic Discourse: Questions on Feminist Strategies in Adrienne Rich*. Amsterdam, Netherlands, and Philadelphia, PA: John Benjamins.

Dixon, Paul B. 1989. "Feedback, Strange Loops and Machado de Assis's 'O espelho.'" *Romance Quarterly* 36.2 (1 May): 213–21.

Dolinin, Alexander. 1995. "Eugene Onegin." In Vladimir E. Alexandrov, ed., *The Garland Companion to Vladimir Nabokov*, 117–30. New York, NY: Garland.

Eco, Umberto. 1979. "The Myth of Superman." In Eco, *The Role of the Reader: Explorations in the Semiotics of Texts*, 107–24. Bloomington, IN: Indiana University Press.

Emerson, Caryl, trans. and ed. 1984. Mikhail Bakhtin, *Problems of Dostoevsky's Poetics*. Theory and History of Literature, vol. 8. Minneapolis, MN: University of Minnesota Press, 1984.

Falen, James E., trans. 1990. Alexander Pushkin, *Eugene Onegin: A Novel in Verse*. New York. NY: Oxford University Press.

Farías, Victor. 1991. *Heidegger and Nazism*. French materials translated by Paul Burrell, with the advice of Dominic Di Bernardi, German materials translated by Gabriel R. Ricci. Edited by Joseph Margolis and Tom Rockmore. Philadelphia, PA: Temple University Press.

Fleck, Ludwig. 1986. "The Problem of Epistemology." In Robert S. Cohen and Thomas Schnelle, eds., *Cognition and Fact: Materials on Ludwig Fleck*, 79–112. Dordrecht, Netherlands: Reidel.

Fonagy, Peter, and Mary Target. 1997. "Playing with Reality: I. Theory of Mind and the Normal Development of Psychic Reality." *International Journal of Psycho-Analysis* 77 (April): 217–33.

Foucault, Michel. 1969/1983. « Qu'est-ce qu'un auteur? » ["What is an Author?"] *littoral* 9 (June): 3–32.

Funkenstein, Amos. 2000. "Theological Interpretations of the Holocaust: A Balance." In Michael L. Morgan, ed., *A Holocaust Reader: Responses to the Nazi Extermination*, 172–82. Oxford, UK, and New York, NY: Oxford University Press.

Gadamer, Hans-Georg. 1960/1975. *Wahrheit und Methode: Grundzüge einer philosophischen Hermeneutik* ["Truth and Method: Main Features of a Philosophical Hermeneutic"]. 4th edition. Tübingen, Germany: Mohr.

Gentzler, Edwin. 2017. *Translation and Rewriting in the Age of Post-Translation Studies*. London, UK, and New York, NY: Routledge.

Göransson, Johannes. 2018. *Transgressive Circulation: Essays on Translation*. Blacksburg, VA: Noemi.

Grundmann, Reiner, and Christos Mantziaris. 1991. "Fundamentalist Intolerance or Civil Disobedience?: Strange Loops in Liberal Theory." *Political Theory* 19.4 (November): 572–605.

Gukasyan, Tatevik, trans. 2019. Natalia S. Avtonomova, "Philosophy, Translation, 'Untranslatability': Cultural and Conceptual Aspects." Translation of Avtonomova unpub. in Spitzer 2019a: 1–35.

Harari, Josué, trans. 1979. Michel Foucault, "What is an Author?" In Harari, ed., *Textual Strategies: Perspectives in Post-structuralist Criticism*, 141–60. Ithaca, NY: Cornell University Press.

Haspelmath, Martin. 1997. *Indefinite Pronouns*. Oxford, UK, and New York, NY: Oxford University Press.
Heidegger, Martin. 1927/1967. *Sein und Zeit* ["Being and Time"]. Tübingen, Germany: Max Niemeyer.
Heidegger, Martin. 1985. *Unterwegs zur Sprache* ["Underway to Language"]. Vol. 12 of the *Gesamtausgabe* ["Collected Works"]. Frankfurt a. M., Germany: Klostermann.
Heidegger, Martin. 2014. *Überlegungen II–VI (Scharze Hefte 1931–1938)* ["Reflections II–VI (Black Notebooks 1931–1938)"]. Edited by Peter Trawny. In Part IV, *Aufzeichnungen* ["Notes and Records"], Vol. 94 of the *Gesamtausgabe* ["Collected Works"]. Frankfurt a. M., Germany: Klostermann.
Heidegger, Martin. 2019a. *Vier Hefte I und II (Schwarze Hefte 1947–1950)* ["Four Notebooks 1 and 2 (Black Notebooks 1947–1950)"]. Edited by Peter Trawny. In Part IV, *Aufzeichnungen* ["Notes and Records"], Vol. 99 of the *Gesamtausgabe* ["Collected Works"]. Frankfurt a. M., Germany: Klostermann.
Heidegger, Martin. 2019b. *Vigiliae und Notturno (Schwarze Hefte 1952/53 bis 1957)*. Edited by Peter Trawny. In Part IV, *Hinweise und Aufzeichnungen* ["Notes and Records"], Vol. 100 of the *Gesamtausgabe* ["Collected Works"]. Frankfurt a. M., Germany: Klostermann.
Hermans, Theo. 1996. "The Translator's Voice in Translated Narrative." *Target* 8.1: 23–48.
Hermans, Theo. 1997. "Translation and Normativity." In Christina Schäffner, ed., *Translation and Norms*, special issue of *Current Issues in Language and Society* 5.1–2: 51–72.
Heyvaert, Stefan, trans. 1992. Antoine Berman, *The Experience of the Foreign: Culture and Translation in Romantic Germany*. Translation of Berman 1984. Albany, NY: SUNY Press.
Hix, H.L., in conversation with Don Mee Choi. 2015. "Don Mee Choi on Kim Hyesoon's *Mommy Must Be a Fountain of Feathers*." *Essay Press Listening Tour* #22, 1–6. Online at http://www.essaypress.org/wp-content/uploads/2015/04/hixltspread.pdf. Accessed January 12, 2019.
Hofstadter, Douglas R. 1979/1989. *Gödel, Escher, Bach: An Eternal Golden Braid*. New York, NY: Basic.
Hofstadter, Douglas R. 1997. *Le Ton Beau de Marot: In Praise of the Music of Language*. New York, NY: Basic.
Hofstadter, Douglas, trans. 1999. *Eugene Onegin: A Novel in Verse by Alexander Sergeevich Pushkin*. New York, NY: Basic.
Hofstadter, Douglas. 2007. *I Am a Strange Loop*. New York, NY: Basic.
Hofstadter, Douglas. 2009. "Translator, Trader: An Essay on the Pleasantly Pervasive Paradoxes of Translation." Published in a back-to-front omnibus with Hofstadter's translation of Françoise Sagan, *That Mad Ache: A Novel*. New York, NY: Basic.

House, Juliane. 2005. "Text and Context in Translation." *Journal of Pragmatics* 38.3 (March): 338–58.

Howard, Richard, trans. 1967. Roland Barthes, "The Death of the Author." *Aspen* 5–6.3: 2–6.

Jaitner, Sabina Folnović. 2019. "Philosophical Untranslatables and the Concept of Equivalence." In Spitzer 2019a: 97–118.

Kankimäki, Mia. 2018. *Naiset joita ajattelen öisin* ["The Women I Think About at Night"]. Helsinki: Otava.

Kohut, Heinz. 1968. "The Psychoanalytic Treatment of Narcissistic Personality Disorders: Outline of a Systematic Approach." *Problems of Psychopathology and Therapy* 23: 86–113.

Kohut, Heinz. 1971. *The Analysis of the Self: A Systematic Approach to the Psychoanalytic Treatment of Narcissistic Personality Disorders*. New York, NY: International Universities Press.

Kohut, Heinz. 1977. *The Restoration of the Self*. New York, NY: International Universities Press.

Kohut, Heinz. 1983. "Selected Problems of Self Psychological Theory." In Joseph D. Lichtenberg and Samuel Kaplan, eds., *Reflections on Self Psychology*, 387–416. Hillsdale, NJ: Analytic.

Kosinski, Jerzy. 1970. *Being There*. New York, NY: Harcourt Brace Jovanovich.

Lacan, Jacques. 1973. *Écrits: A Selection*. Translated by Alan Sheridan. London, UK: Cassell.

Ladmiral, Jean-René. 1986. "Sourciers et ciblistes" ["Sourciers and Targetists"]. *Revue d'esthétique* 12: 33–42.

Ladmiral, Jean-René. 1994. *Traduire: théorèmes pour la traduction* ["To Translate: Theorems About Translation"]. Paris, France: Gallimard.

Ladmiral, Jean-René. 2000. "Traduire des philosophes" ["To Translate the Philosophers"]. In Jacques Moutaux and Oliver Bloch, eds., *Traduire les philosophes*, 49–73. Paris, France: Éditions de la Sorbonne.

Launer, John. 2010. "Double Binds and Strange Loops." *The Fellowship of Postgraduate Medicine: Postgraduate Medical Journal* 86.1011 (January): 63.

Le Moigne, Jean-Louis, and Edgar H. Sibley. 1986. "Information—Organization—Decision: Some Strange Loops." *Information & Management* 11.5: 237–44.

Lefevere, André. 1992. *Translation, Rewriting, and the Manipulation of Literary Fame*. London, UK, and New York, NY: Routledge.

Littau, Karin. 1993. "Intertextuality and Translation: *The Waste Land* in French and German." In Catriona Picken, ed., *Translation: The Vital Link*, 63–9. London, UK: Chameleon.

Littau, Karin. 1997. "Translation in the Age of Postmodern Production: From Text to Intertext to Hypertext." *Forum for Modern Language Studies* 33: 81–96.

Littler, Margaret. 2013. "Strange Loops and Quantum Turns in Barbara Köhler's *Niemands Frau*." *German Monitor* 78: 163–84.

Liu, Lydia H. 1995. *Translingual Practice: Literature, National Culture, and Translated Modernity: China, 1900–1937*. Stanford, CA: Stanford University Press.

Liu, Lydia H., ed. 1999. *Tokens of Exchange: The Problem of Translation in Global Circulations*. Durham, NC: Duke University Press.

Lotbinière-Harwood, Susanne de. 1991. *Re-Belle et Infidèle: la traduction comme pratique de réécriture au féminin = The Body Bilingual, Translation as Rewriting in the Feminine*. Toronto, ON: Women's Press.

Luther, Martin. 1530/2004. "Ein Sendbrief vom Dolmetschen" ["A Circular Letter on Translation"]. *Die Übersetzung* ["Translation"], Appendix A. Alois Payer, ed., *Einführung in die Exegese von Sanskrittexten* ["Introduction to the Exegesis of Sanskrit Texts"], chapter 4. http://www.payer.de/exegese/exeg04a.htm. Accessed January 8, 2019.

Macquarrie, John, and Edward Robinson, trans. 1962. Martin Heidegger, *Being and Time*. Translation of Heidegger 1927/1967. Oxford, UK: Blackwell.

Marais, Kobus. 2014. *Translation Theory and Development Studies: A Complexity Theory Approach*. London, UK, and New York, NY: Routledge.

Massumi, Brian, trans. 1987. Gilles Deleuze and Félix Guattari, *A Thousand Plateaus*. Translation of Deleuze and Guattari 1980, *Capitalism and Schizophrenia*, vol. 2. Minneapolis, MN: University of Minnesota Press.

Massumi, Brian. 2002. *Parables for the Virtual: Movement, Affect, Sensation*. Durham, NC: Duke University Press.

McSweeney, Joyelle. 2014. "What is the Necropastoral?" Harriet Blog, April 29. Online at https://www.poetryfoundation.org/harriet/2014/04/what-is-the-necropastoral. Accessed January 12, 2019.

Meek, Mary Elizabeth, trans. 1971. Émile Benveniste, *Problems in General Linguistics*. Translation of Benveniste 1966. Coral Gables, FL: University of Miami Press.

Mensah, Patrick, trans. 1996. Jacques Derrida, *Monolingualism of the Other; or, the Prosthesis of Origin*. Translation of Derrida 1996. Stanford, CA: Stanford University Press.

Merleau-Ponty, Maurice. 1945/1970. *Phenomenology of Perception*. Translated by Colin Smith. London, UK: Routledge and Kegan Paul.

Meschonnic, Henri, trans. 1970. *Les Cinq Rouleaux (Le Chant des chants, Ruth, Comme ou les Lamentations, Parole du sage, Esther)* ["The Five Scrolls (The Song of Songs, Ruth, As or the Lamentations, Word of the Wise, Esther"]. Paris, France: Gallimard.

Meschonnic, Henri, trans. 1981. *Jona et le signifiant errant* ["Jonah and the Wandering Signifier"]. Paris, France: Gallimard.

Meschonnic, Henri. 1982. *Critique du rhythme: Anthropologie historique du langage* ["Critique of Rhythm: Historical Anthropology of Language"]. Paris, France: Verdier.

Meschonnic, Henri, trans. 2001. *Gloires, traduction des psaumes* ["Glories, Translation of the Psalms"]. Paris, France: Desclée de Brouwer.

Meschonnic, Henri, trans. 2002. *Au commencement, traduction de La Genèse* ["In the Beginning, Translation of Genesis"]. Paris, France: Desclée de Brouwer.

Meschonnic, Henri, trans. 2003. *Les Noms, traduction de L'Exode* ["The Names, Translation of Exodus"]. Paris, France: Desclée de Brouwer.

Meschonnic, Henri, trans. 2005. *Et il appelé, traduction du Lévitique* ["And He Called, Translation of Leviticus"]. Paris, France: Desclée de Brouwer.

Meschonnic, Henri. 2007. *Ethique et politique du traduire* ["Ethics and Politics of Translating"]. Paris, France: Verdier.

Moore, Rich, director. 2012. *Wreck-it Ralph*. Burbank, CA: Walt Disney Animation Studios.

Morden, Michael. 1990. "Free Will, Self-Causation, and Strange Loops." *Australasian Journal of Philosophy* 68.1 (1 March): 59–73.

Morton, Timothy. 2007. *Ecology Without Nature*. Cambridge, MA: Harvard University Press.

Munday, Jeremy. 2001/2012. *Introducing Translation Studies: Theories and Applications*. 33rd edition. London, UK, and New York, NY: Routledge.

Munday, Jeremy, ed. 2007. *Translation as Intervention*. London, UK, and New York, NY: Continuum.

Nabokov, Vladimir. 1941. "The Art of Translation." *New Republic* (August 5). Online at https://newrepublic.com/article/62610/the-art-translation. Accessed September 27, 2019.

Nabokov, Vladimir. 1955. "Problems of Translation: Onegin in English." *Partisan Review* 22: 511–12.

Nabokov, Vladimir, trans. 1965. Aleksandr Sergeevich Pushkin, *Eugene Onegin: A Novel in Verse*. 2 vols. Princeton, NJ: Bollingen Foundation.

Nida, Eugene A. 1964. *Toward a Science of Translating: With Special References to Principles and Procedures of Bible Translating*. Leiden, Netherlands: Brill.

Nida, Eugene A., and Charles R. Taber. 1969. *Theory and Practice of Translation*. Leiden, Netherlands: Brill.

Nida, Eugene A., and Jan de Waard. 1986. *From One Language to Another: Functional Equivalence in Bible Translating*. Nashville, TN, Camden, NJ, and New York, NY: Thomas Nelson.

Niranjana, Tejaswini. 1992. *Siting Translation: History, Poststructuralism, and the Colonial Context*. Berkeley and Los Angeles, CA: University of California Press.

Pagano, Adriana, and Maria Lúcia Vasconcellos. 2003. "Estudos da tradução no Brasil: reflexões sobre teses e dissertações elaboradas por pesquisadores brasileiros nas décadas de 1980 e 1990" ["Translation Studies in Brazil: Reflections on Theses and Dissertations Prepared by Brazilian Researchers in the 1980s and 1990s"]. *DELTA: Documentação de Estudos em Linguística Teórica e Aplicada* 19: 1–25.

Palombo, Joseph. "Narratives, Self-Cohesion, and the Patient's Search for Meaning." *Clinical Social Work Journal* 20 (1992): 249–70.

Peirce, Charles S. 1931–58. *Collected Papers of Charles Sanders Peirce*. 8 vols. Vols. 1–6 edited by Charles Hartshorne and Paul Weiss; vols. 7–8 edited by Arthur W. Burks. Cambridge, MA: Harvard University Press.

Peirce, Charles S. 1992–98. *The Essential Peirce: Selected Philosophical Writings*. 2 vols. Edited by the Peirce Edition Project. Bloomington, IN: Indiana University Press.

Perelman, Chaim. 1982. *The Realm of Rhetoric*. Translated by William Kluback. Notre Dame, IN: University of Notre Dame Press.

Perelman, Chaim, and L. Olbrechts-Tyteca. 1969. *The New Rhetoric: A Treatise on Argumentation*. Translated by John Wilkinson and Purcell Weaver. Notre Dame, IN: University of Notre Dame Press.

Peters, John Durham. 2001. *Speaking Into the Air: A History of the Idea of Communication*. Chicago, IL: University of Chicago Press.

Pym, Anthony. 1993. "Performatives as a Key to Modes of Translational Discourse." In János Kohn and Klinga Klaudy, eds., *Transferre necesse est ... Current Issues in Translation Theory*, 47–62. Szombathely, Hungary: Pädagogische Hochschule Berzsenyi Dániel.

Pym, Anthony. 2010. *Exploring Translation Theories*. London, UK, and New York, NY: Routledge.

Razumnaya, Anna. 2012. "Onegin in English: Against Nabokov." *Literary Imagination* 14.3: 277–91.

Reimer, Georg, ed. 1860. *Aus Schleiermachers Leben. In Briefe* ["From Schleiermacher's Life: in Letters"]. Vol. 2, *Von Schleiermacher's Anstellung in Halle—October 1804—bis an sein Lebensende—den 12. Februar 1834* ["From Schleiermacher's Placement in Halle, October 1804, till the end of his life, February 12, 1834"]. Berlin: Perthes.

Richards, I.A. 1956. *Practical Criticism*. New York, NY: Harvest.

Ricoeur, Paul. 2004. *Sur la traduction: Grandes difficultés et petits bonheurs de la traduction* ["On Translation: Great Difficulties and Small Pleasures of Translation"]. Paris, France: Bayard.

Robinson, Douglas. 1985. "Intentions, Signs, and Interpretations: C. S. Peirce and the Dialogic of Pragmatism." *Kodikas/Code/Ars Semeiotica* 8.3-4: 179–93.

Robinson, Douglas. 1991. *The Translator's Turn*. Baltimore, MD, and London, UK: Johns Hopkins University Press.

Robinson, Douglas. 1992. *Ring Lardner and the Other*. New York, NY: Oxford University Press.

Robinson, Douglas, trans. 1993a. Aleksis Kivi, *Heath Cobblers (Nummisuutarit) and Kullervo*. St. Cloud, MN: North Star Press of St. Cloud.

Robinson, Douglas, trans. 1993b. Martin Buber, "On the Diction of a German Translation of the Scripture." *Translation and Literature* 2: 105–10.

Robinson, Douglas. 1995. "Translation and the Double Bind." *Studies in the Humanities* 22.1/2 (December): 1–11.

Robinson, Douglas, ed. 1997/2014. *Western Translation Theory from Herodotus to Nietzsche*, 225–38. 3rd edition. London, UK, and New York, NY: Routledge.

Robinson, Douglas. 1997/2020. *Becoming a Translator: An Introduction to the Theory and Practice of Translation*. 4th revised edition. London, UK, and New York, NY: Routledge.

Robinson, Douglas. 1997a. "Looking Through Translation: A Response to Gideon Toury and Theo Hermans." In Christina Schäffner, ed., *Translation and Norms*, special issue of *Current Issues in Language and Society* 5.1–2: 114–23.

Robinson, Douglas. 1997b. *What Is Translation? Centrifugal Theories, Critical Interventions*. Kent, OH: Kent State University Press.

Robinson, Douglas. 1998. "Kugelmass, Translator (Some Thoughts on Translation and its Teaching)." Peter Bush and Kirsten Malmkjaer, eds., *Rimbaud's Rainbow: Literary Translation in Higher Education*, 47–61. Amsterdam, Netherlands, and Philadelphia, PA: John Benjamins.

Robinson, Douglas. 2001. *Who Translates? Translator Subjectivities Beyond Reason*. Albany, NY: SUNY Press.

Robinson, Douglas. 2003. *Performative Linguistics: Speaking and Translating as Doing Things With Words*. London, UK, and New York, NY: Routledge.

Robinson, Douglas. 2006. "Translation and the Nationalist/Migrant Double Bind." *TIS: Translation and Interpreting Studies* 1.1 (Spring): 110–23.

Robinson, Douglas. 2008. *Estrangement and the Somatics of Literature: Tolstoy, Shklovsky, Brecht*. Baltimore, MD, and London, UK: Johns Hopkins University Press.

Robinson, Douglas. 2009. "Adding a Voice or Two: Translating Pentti Saarikoski for a Novel." In B.J. Epstein, ed., *Northern Lights: Translation in the Nordic Countries*, 213–38. Bern, Switzerland: Peter Lang.

Robinson, Douglas. 2010. "Liar Paradox Monism: A Wildean Solution to the Explanatory Gap between Materialism and Qualia." *Minerva* 14: 66–106. Online at http://www.minerva.mic.ul.ie/Vol14/Monism.pdf. Accessed July 1, 2021.

Robinson, Douglas. 2011. *Translation and the Problem of Sway*. Amsterdam, Netherlands, and Philadelphia, PA: John Benjamins.

Robinson, Douglas. 2012a. *First-Year Writing and the Somatic Exchange*. New York, NY: Hampton Press.

Robinson, Douglas. 2012b. "Rhythm as Knowledge-Translation, Knowledge as Rhythm-Translation." *Global Media Journal—Canadian Edition* 5.1: 75–94.

Robinson, Douglas. 2013a. *Feeling Extended: Sociality as Extended Body-Becoming-Mind*. Cambridge, MA: MIT Press.

Robinson, Douglas. 2013b. *Schleiermacher's Icoses: Social Ecologies of the Different Methods of Translating*. Bucharest, Romania: Zeta Books.

Robinson, Douglas. 2014a. "Embodied Translation: Henri Meschonnic on Translating For/Through the Ear and the Mouth." *Parallèles* 26: 38–52. Online at http://www.paralleles.unige.ch/tous-les-numeros/numero-26/robinson.html. Accessed July 1, 2021.

Robinson, Douglas. 2014b. "The Inscience of Translation." *International Journal of Society, Culture, and Language* 2.2 (Fall): 25–40. Online at http://ijscl.net/article_5432_848.html. Accessed July 1, 2021.

Robinson, Douglas. 2014c. "Problems in Translating 'Circulatory' Terms from Aristotle's Greek and Mencius's Chinese: *pistis* 'persuading/being persuaded' and 治 *zhì* 'governing/being governed' in English." In Laurence K.P. Wong, John C.Y. Wang, and Chan Sin-wai, eds., *Two Voices in One*, 159–74. Newcastle upon Tyne, UK: Cambridge Scholars.

Robinson, Douglas. 2015. *The Dao of Translation: An East–West Dialogue*. London, UK, and Singapore: Routledge.

Robinson, Douglas. 2016a. "Benveniste and the Periperformative Structure of the Pragmeme." In Keith Allen, Alessandro Capone, Istvan Kecskes, and Jacob L. Mey, eds., *Pragmemes and Theories of Language Use*, 85–104. Dordrecht, Netherlands: Springer.

Robinson, Douglas. 2016b. *The Deep Ecology of Rhetoric in Mencius and Aristotle*. Albany, NY: SUNY Press.

Robinson, Douglas. 2016c. "Pushing-Hands and Periperformativity." In Douglas Robinson, ed., *The Pushing Hands of Translation and its Theory: In Memoriam Martha Cheung, 1953–2013*, 173–216. London, UK, and Singapore: Routledge.

Robinson, Douglas. 2016d. *Semiotranslating Peirce*. Tartu, Estonia: Tartu Library of Semiotics.

Robinson, Douglas, trans. 2017a. Aleksis Kivi, *The Brothers Seven*. Bucharest, Romania: Zeta Books.

Robinson, Douglas. 2017b. *Aleksis Kivi and/as World Literature*. Leiden, Netherlands, and Boston, MA: Brill.

Robinson, Douglas. 2017c. *Critical Translation Studies*. London, UK, and Singapore: Routledge.

Robinson, Douglas. 2017d. *Exorcising Translation: Towards an Intercivilizational Turn*. New York, NY: Bloomsbury Academic.

Robinson, Douglas. 2017e. *Translationality: Essays in the Translational-Medical Humanities*. London, UK, and Singapore: Routledge.

Robinson, Douglas. 2019. *Transgender, Translation, Translingual Address*. New York, NY: Bloomsbury Academic.

Robinson, Douglas. 2020a. "'A Man Carries His Door': Affective Displacement and Refugee Poetry." In Emma Cox, Sam Durrant, David Farrier, Lyndsey Stonebridge, and Agnes Woolley, eds., *Refugee Imaginaries: Research Across the Humanities*, 248–63. Edinburgh, UK: Edinburgh University Press.

Robinson, Douglas, trans. 2020b. Mia Kankimäki, *The Women I Think About at Night*. New York, NY: Simon & Schuster.

Robinson, Douglas. 2020c. "Reframing Translational Norm Theory Through 4EA Cognition." *Translation, Cognition, Behavior* 3.1: 122–43.

Robinson, Douglas, trans. 2020d. Volter Kilpi, *Gulliver's Voyage to Phantomimia*. Bucharest, Romania: Zeta Books.

Robinson, Douglas. 2021b. "Sixteen Avant-Garde Perspectives on World Literature and the Translator's (In)visibility." *Asymptote* (January). Online at https://www.asymptotejournal.com/special-feature/sixteen-avantgarde-perspectives-on-world-literature-and-the-translators-invisibility-douglas-robinson/. Accessed July 1, 2021.

Robinson, Douglas. 2022a. *The Behavioral Economics of Translation*. London, UK, and New York, NY: Routledge.

Robinson, Douglas. 2022b. *Translating the Monster: Volter Kilpi in Orbit Beyond (Un-)translatability*. Leiden, Netherlands, and Boston, MA: Brill.

Robinson, Douglas. 2023. *Translation as a Form: A Centennial Commentary on Walter Benjamin's "The Task of the Translator."* London, UK, and New York, NY: Routledge.

Robinson, Douglas. Unpub. "Kääntämisen kääntöpiirit/The Tropics of Translation." Photocopy.

Rosengrant, Judson. 1994. "Nabokov's Theory and Practice of Translation: 1941–1975." *Slavic and East European Journal* 381: 29–32.

Sakai Naoki. 1997. *Translation and Subjectivity: On "Japan" and Cultural Nationalism*. Minneapolis, MN: University of Minnesota Press.

Schiavi, Giuliana. 1996. "There Is Always a Teller in a Tale." *Target* 8.1: 1–21.

Schlegel, August Wilhelm von. 1791/1962. "Dante—Über die göttliche Komödie" ["Dante—On the *Divine Comedy*"]. In *Sprach und Poetik* ["Speech and Poetics"], vol. 1 of Edgar Lohner, ed., *Schlegels Kritische Schriften und Briefe* ["Schlegel's Critical Writings and Letters"], 67–87. Stuttgart, Germany: Kohlhammer.

Schleiermacher, Friedrich. 1813/2002. "Ueber die verschiedenen Methoden des Übersetzens" ["On the Different Methods of Translating"]. In Martin Rößler, with the assistance of Lars Emersleben, eds., *Akademievorträge* ["Academy Addresses"], 67–93. Part I of *Schriften und Entwürfe* ["Writings and Drafts"]. Vol. 11 of *Kritische Gesamtausgabe* ["Critical Complete Works"]. Berlin, Germany: BBAW.

Schore, Allan N. 1994. *Affect Regulation and the Origin of the Self: The Neurobiology of Emotional Development*. Mahwah, NJ: Erlbaum, 1994.

Schore, Allan N. 2003a. *Affect Regulation and the Repair of the Self*. New York. NY: Norton.

Schore, Allan N. 2003b. *Affect Dysregulation and Disorders of the Self*. New York, NY: Norton.

Sedgwick, Eve. 2003. *Touching Feeling: Affect, Pedagogy, Performativity*. Durham, NC: Duke University Press.

Short, T.L. 2004. "The Development of Peirce's Theory of Signs." In Cheryl Misak, ed., *The Cambridge Companion to Peirce*, 214–40. Cambridge, UK: Cambridge University Press.

Short, T.L. 2007. *Peirce's Theory of Signs*. Cambridge, UK: Cambridge University Press.

Simeoni, Daniel. 1998. "The Pivotal Status of the Translator's Habitus." *Target* 10.1: 1–39.

Simon, Sherry. 1995. *Gender in Translation: Cultural Identity and the Politics of Transmission*. London, UK, and New York, NY: Routledge.

Solms, Mark. 1996. "Towards an Anatomy of the Unconscious." *Journal of Clinical Psychoanalysis* 5.3: 331–67.

Spitzer, D.M., ed. 2019a. *Philosophy's Treason: Studies in Translation and Philosophy*. Wilmington, DE, and Malaga, Spain: Vernon.

Spitzer, D.M. 2019b. "Introduction: Philosophy's Treason." In Spitzer 2019a: ii–xxiv.

Spivak, Gayatri Chakravorty, trans. 1976. Jacques Derrida, *Of Grammatology*. Translation of Derrida 1967. Baltimore, MD, and London, UK: Johns Hopkins University Press.

Stambaugh, Joan, trans. 1996. Martin Heidegger, *Being and Time*. Translation of Heidegger 1927/1967. Albany, NY: SUNY Press.

Steiner, George. 1975/1998. *After Babel: Aspects of Language and Translation*. Oxford, UK, and New York, NY: Oxford University Press.

Stern, Daniel. 1977. *The First Relationship*. Cambridge, MA: Harvard University Press.

Stern, Daniel N., Roanne K. Barnett, and Susan Spieker. 1983. "Early Transmission of Affect: Some Research Issues." In Justin D. Call, Eleanor Galenson, and Robert L. Tyson, eds., *Frontiers of Infant Psychiatry*, 52–69. New York, NY: Basic.

Taylor, Graeme J. 1987. *Psychosomatic Medicine and Contemporary Psychoanalysis*. Madison, CT: International Universities Press, 1987.

Tekwa, Kizito. 2019. "Circumventing Literalness Through a Consecutive 'Interpretranslation' Training Method." Second WITTA International Symposium on Translation Education, Kunming, Yunnan Province, PRC, July 4.

Thandeka. 2005. "Schleiermacher's Affekt Theology." *International Journal of Practical Theology* 9.2: 197–216.

Tiffany, Daniel. 1998. *Radio Corpse: Imagism and the Cryptoaesthetic of Ezra Pound*. Cambridge, MA: Harvard University Press.

Tiffany, Daniel. 2014. *My Silver Planet: A Secret History of Poetry and Kitsch*. Baltimore, MD, and London, UK: Johns Hopkins University Press.

Toury, Gideon. 1980. *In Search of a Theory of Translation*. Tel Aviv, Israel: Porter Institute for Poetics and Semiotics, Tel Aviv University.

Toury, Gideon. 1995. *Descriptive Translation Studies and Beyond*. Amsterdam, Netherlands, and Philadelphia, PA: John Benjamins.

Toury, Gideon. 1997. "A Handful of Paragraphs on 'Translation' and 'Norms.'" In Christina Schäffner, ed., *Translation and Norms*, special issue of *Current Issues in Language and Society* 5.1–2: 10–32.

Trubikhina, Julia. 2008. "The Metaphysical 'Affinity of the Unlike': Strategies of Nabokov's Literalism." *Intertexts* 12.1: 55–73.

Tymoczko, Maria. 2007/2010. *Enlarging Translation, Empowering Translators*. Manchester, UK, and Kinderhook, NY: St. Jerome.
Venuti, Lawrence. 1995. *The Translator's Invisibility*. London, UK, and New York, NY: Routledge.
Venuti, Lawrence, ed. 1998. *Translation and Minority*. Special issue of *The Translator* 4.2. Manchester, UK: St. Jerome.
Vihelmaa, Ella. 2018. "Kielen kääntöpuolella: Kuinka tutkia toislajisten merkkien kääntymistä ihmiskielelle?" ["On the Animal Side of Language: How to Study the Translation of Nonhuman Signs into Human Language?"]. Licentiate thesis, University of Eastern Finland. Online at epublications.uef.fi/pub/urn_nbn_fi_uef-20190920/urn_nbn_fi_uef-20190920.pdf. Accessed August 20, 2019.
Weinberg, Igor. 2000. "The Prisoners of Despair: Right Hemisphere Deficiency and Suicide." *Neuroscience and Biobehavioral Reviews* 24.8: 799–815.
Weinsheimer, Joel, and Donald G. Marshall, trans. 2004. Hans-Georg Gadamer, *Truth and Method*. Translation of Gadamer 1960/1975. 2nd revised edition. New York, NY: Continuum.
Wenger, Etienne. 1998. *Communities of Practice: Learning, Meaning, and Identity*. Cambridge, UK: Cambridge University Press.
Williams, Nicholas Morrow. 2018. "Tropes of Entanglement and Strange Loops in the 'Nine Avowals' of the Chuci." *Bulletin of the School of Oriental and African Studies* 81.2: 277–300.
Williford, Ken, and Uriah Kriegel, eds. 2006. *Self-Representational Approaches to Consciousness*. Cambridge, MA: MIT/Bradford.
Winnicott, D.W. 1958. *Through Paediatrics to Psycho-Analysis*. New York, NY: Basic.
Wolf, Ernest S. 1988. *Treating the Self: Elements of Clinical Self Psychology*. New York, NY: Guilford.
Zukofsky, Celia T. and Louis, trans. 1969. *Catullus (Gai Valeri Catulli Veronensis liber)*. London, UK: Cape Goliard.

Index

"'A Man Carries His Door'" (Robinson) 191
"Adding a Voice or Two" (Robinson) 69
Affect 141, 143–4; alienation-like (Schleiermacher) 59; -becoming-cognition 141–2, 158; (dys)regulation of (Schore) 170, 172, 174, 176; joyless (Hofstadter) 52; and "locking-in" 76, 138, 156–7, 175; in Meschonnic 152, 154–5, 158; mirroring of 171; misattunement of 172–3; regulation of 171–3, 176; as shaping/guiding cognition 26; suppressed by the KMP 177; theory 24, 160
Affect Dysregulation and Disorders of the Self (Schore) 170
Affect Regulation and the Origin of the Self (Schore) 170
Affect Regulation and the Repair of the Self (Schore) 170
After Babel (Steiner) 112
Against World Literature (Apter) 110, 117
Agency 163; activist 161–2; heroic/super 159, 163, 165; lack of 103, 105–6, 160; projected onto the foreign 62; shared 52; uncertain 106–7, 117, 123
Aleksis Kivi and/as World Literature (Robinson) 19, 188, 200n1
Alienation 59; *see* Entfremdung
Allgemeine Äquivalent (Marx) 133; *see* General equivalent
Analogy 2, 6, 15, 55, 169, 172, 175, 188; and Capgras as serial retranslation 67; in Hofstadter 9–10, 69; in the EGP 197n5; in Schleiermacher 55–9, 63–5; in Meschonnic 149; in Luther 12; *see also* Level-shifts
Andrew, Sally 167
"Anonymous, voiceless, invisible translator slaving away" (Marais) 161, 164–5

Aporia 24, 102–5
Apter, Emily (EAP) 67, 110, 117
Aristotle (AP) 6, 103–4, 122, 142, 155, 192n2
Arndt, Ernst Moritz 65
Arrojo, Rosemary (RAP) 23, 70–2, 77, 79, 84–6, 94–5, 197n1, 199n2
Ashby, Hal 111
Attachment theory (Bowlby) 170
Audience-effects: of the "I," 6–7, 14, 18, 23, 53–4, 69, 71, 180; anti-hoaxical 15; of the aporia 104; of archaizing equivalence 16–17, 21; culture-constitutive 7–8; periperformative, of identity 127, 129; reciprocal 48; of the regime of homolingual address 135
Augustine 40
Austin, J.L. 24, 106
Author-function (Foucault) 23, 42, 70–3, 77, 82, 85–7, 93–4; of Adrienne Rich 72–3, 78, 83; of the incoherent source author 14; of Luther 12–3; as regulatory 94–5
"Auxiliary cortex" (Diamond et al.) 171
Aviram, Amittai (AAP) 146–50
Avtonomova, Natalia S. (NSAP) 25, 110, 116–27, 132

Bach, Johann Sebastian 184
Baker, Mona 49, 107, 158
Bakhtin, Mikhail 9, 69, 77
Ballistic word demons (Dennett) 4, 18, 41, 44, 51
Barthes, Roland (RBP) 24, 71, 84, 86–7, 91–2, 99
Bassnett, Susan 199n2
Bataille, Georges 185
Becoming a Translator (Robinson) 27, 29, 37–8, 132
Behavioral Economics of Translation, The (Robinson) 54, 108, 196n1

Being and Time (Heidegger/Macquarrie/ Robinson) 110
Being There (Kosinski/Ashby) 111
Benjamin, Walter (WBP) 16
Benveniste, Emile (EBP 2) 146–8
"Benveniste and the Periperformative Structure of the Pragmeme" (Robinson) 103
Berg, Aase (AaBP) 181
Berman, Antoine (ABP) 116, 124–6, 199n4
Bernstein, Charles 146
Beyonsense 196n1; *see Zaum*
Bible 16–7; Hebrew 26, 150, 153; Vulgate (Jerome) 41
Blattner, William 198n1
Blum-Kulka, Shoshana 32
Body-becoming-mind 148, 157
Body-in-language continuum (Meschonnic) 151, 153–5
Boisacq, Emile (EBP 1) 146
Borchardt, Rudolf 17
Bowker, Lynne 35
Bowlby, John 170
Brecht, Bertolt 56
Brothers Seven, The (Kivi/Robinson) 19
Buber, Martin 153
Burke, Kenneth 157
Butler, Judith 182

Cannibalism, translation as (de Campos) 184, 186
Capgras delusion 66–8
Capitalized Essence of Consciousness (Hofstadter) 183
Cassin, Barbara 110
Catcher in the Rye (Salinger) 69–70
Cervantes Saavedra, Miguel de 188
Chain of signifiers 90
Chamade, La (Sagan) 2, 40, 49, 70
Chamberlain, Lori 199n2
Choi, Don Mee (DMCP) 188–90
Chomsky, Noam (NCP) 33
Cicero, Marcus Tullius (MTCP) 31, 38, 41
"Circular Letter on Translation" (Luther) 11–13, 42
Circular verbs 27–8

Cofiguration (Sakai) 26; intercivilizational 108; schema of 133
Cognitive schema, cyclical 87–91
Cognitive science 32–3, 37, 51, 56, 86
Commens (Peirce) 142–4
Communities of Practice (Wenger) 146, 154
Complexity theory 166
Conation 143, 155
Congenital analgesiacs 168
Congenital sensory neuropaths 168
Cosmopolitanism 58, 60–5
Critical Translation Studies 119, 128, 130
Critical Translation Studies (Robinson) 8, 27, 109, 119

Damasio, Antonio 141, 170
Dante Alighieri (DAP) 6–7, 17, 188
Dantes Commedia Deutsch (Borchardt) 17
das Man (Heidegger) 25, 111–6, 198n1
Dasein (Heidegger) 25, 110–6
de Campos, Haroldo 184, 186
"Death of the Author, The" (Barthes) 24
Death on the Limpopo (Andrew) 167
"'Death' of the Author and the Limits of the Translator's Invisibility" (Arrojo) 70–1
Deep Ecology of Rhetoric in Mencius and Aristotle, The (Robinson) 27, 153, 155–6, 192n2, 200n3
Deleuze, Gilles 187–9, 200n1
Demons:
—pandemonial (Dennett) 3–4, 14; of the ASP's sources 171; ballistic 4, 18, 41, 44, 51; of the commonsensical reader 142; of the DHP as disbelievers in affect 76, 83, 138–40, 181; of the DHP as theorist of marital love 44, 52, 69, 144–5, 184; of the DHP as theorist of strange loops 44, 50–1, 53, 75, 83, 91, 191; of the DHP as translation theorist 22, 31, 36, 38–9, 47, 52; of the DHP as translator 7, 9, 13, 32, 70; of the DRP 8, 21, 28, 47, 121, 139–41, 188; of the EHSP 22, 40; of the FSP 23, 53–4, 56–9,

62–5; of Holden Caulfield 69–70; of the JDP and the PMP 103; of the JGP 182; of the KMP 160, 162; of the LHLP 131; of the MDDP 82–3; of the NSAP 117; of the RBP 99; of the SNP 135; shared 44–53, 139–40; of the TGP 122; of the THP 97; in translation 14–5, 18–19, 21, 45–6, 54, 125, 174, 196n3; transminoritizing 188–9; —Christian 169; and the demonization-and-purification loop 182, 186
Denaturalization 56
Dennett, Daniel (DDP) 3–4, 50–1, 145, 180
Derrida, Jacques (JDP) 24, 133, 141; and aporias 102–5; and the (ex)orbitant 86–91
Descriptive Translation Studies (DTS) 73, 80, 100
Development studies 108, 161
Díaz-Diocaretz, Miriam (MDDP) 23–4, 70–84, 197n1
Dictionary of Untranslatables (Cassin/Apter et al.) 110, 117
Disciplinary identity-(de)formations 108, 127
Domesticating translation 63–4, 200n1
Double-binds of translation (Robinson) 27; in Heidegger 116
Downward causation 170
Doxicosis (Robinson) 156
"Drawing Hands" (Escher) 13–4, 179
Dyads: I-you 25, 101, 110, 128; mother-child 171–3, 176
Dynamic equivalence (Nida) 54

Echo Maker, The (Powers) 66–8
Eco, Umberto (UEP) 16, 198n2
Ecosis (Robinson) 155
Ein Land, Ein Volk, Eine Sprache 64
Embodiment 26, 111, 137–8, 144, 158, 179; as intersubjective/communalized 146, 148, 156; of discourse 152; rhythms of 148; and the sinthome 182; social action as 154; *see also* Communities of Practice, Knowledge-translation, "Meat-infatuated philosophers"
Emergentism 181
Emotional-energetic-logical interpretant (Peirce) 142–3
Empathy 50, 61
Endless semiosis (Peirce) 90, 197n2
Entelechy (Aristotle) 142–3, 155
Entfremdung 59; *see also* Alienation
Éperons/Spurs (Derrida) 141
Equivalence 22; dynamic (Nida) 54; general (Marx/Moore/Aveling) 133; presumed (Toury) 15, 96; *see Allgemeine Äquivalent*
Erfahrung des fremden, die (Hölderlin/Heidegger) 125, 199n4
Error analysis 121
Escher, M.C. (MCEP) 13–4, 87, 179
Estrangement 59; *see also Verfremdung*
Estrangement and the Somatics of Literature (Robinson) 56
Ethecosis (Robinson) 155
Ethnocentrism 167
Ethnopoetics 38
Ēthos (Aristotle) 6, 103, 122, 192; convergence with *pathos* 104, 123; and *pistis* 123
Eugene Onegin (Pushkin/Nabokov/Hofstadter/Falen) 2, 9, 148, 193-6n5
"Exoteric circle" (Fleck) 108
Experience of the Foreign, The (Hölderlin/Berman/Heyvaert) 125, 199n4; *see Erfahrung des fremden, L'épreuve de l'étranger*
Exploring Translation Theories (Pym) 19, 199n3

Farias, Victor 112
Feeling 36–7, 46, 141, 173–4, 176, 196n3; of the aporia 102; -becoming-thinking 141–4; driving locking-in 76, 138, 156–7, 175; of the Foreign (Schleiermacher) 53–4, 59–63, 65, 124–6, 199n4; rejected by Hofstadter 36, 53, 76, 138–9, 192n1, 197n4; shared 126; as split off from thought 141–3

Feeling Extended (Robinson) 27, 44
"Feminist, 'Orgasmic' Theories of Translation and their Contradictions" (Arrojo) 199n2
Feminist translation scholars 112
Fichte, Johann Gottlieb 65
First Relationship, The (Stern) 171
First-Year Writing and the Somatic Exchange (Robinson) 157
Foreign bodies, as kitsch 185
Foreignizing translation 53, 63, 116, 124, 126
Foucault, Michel (MFP) 23–4, 42, 70–4, 78–80, 84–7, 91–4, 197n1
Freud, Sigmund 133
Friedrich Wilhelm III, King 57–8
FSP demons: cosmopolitan 58; panicky/hysterical 56; nationalist 64; *see* Demons, Schleiermacher
Fusion: of horizons (Gadamer) 124; of marital pandemonia (Hofstadter) 197n4

Gadamer, Hans-Georg 124
Gefühl des fremden (Schleiermacher) 53, 59, 124, 126, 199n4; *see also* Feeling
"Gender and the Metaphorics of Translation" (Chamberlain) 199n2
Gender in Translation (Simon) 199n2
General equivalent (Marx) 133; *see Allgemeine Äquivalent*
Gentzler, Edwin 66, 197n5
Glitch loop (Hofstadter/Liu) 131–2, 135
Gödel, Kurt 1–2, 137, 180, 184
Gödel, Escher, Bach (Hofstadter) 1, 5, 10, 21, 28, 100
Goethe, Johann Wolfgang von 65
Google Translate 23, 33–6, 60
Göransson, Johannes (JGP) 27, 51, 179–91
Gospels 40–2
Grammatologie, De la (Derrida) 87
Grammatology, Of (Derrida/Spivak) 24, 87–8
Guattari, Félix 187–9
Gukasyan, Tatevik (TGP) 25, 110, 116–27
Gulliver's Voyage to Phantomimia (Kilpi/Robinson) 193n4

Habitualization 37–8, 132
Hallucinations 180, 187
Hamlet (Shakespeare) 197n5
Heath Cobblers (Kivi/Robinson) 188
Heidegger, Martin (MHP) 25, 67, 109–16, 147
Heidegger's Being and Time (Blattner) 198n1
Herder, Johann Gottfried (JGHP) 199n4
Hermans, Theo (THP) 23–4, 49, 71, 94–100, 107, 176
Hermeneutic circle 27
Hermeneutics 108, 157
Hertz, Neil 106
Heterolingual address, attitude of (Sakai) 128–9, 135, 146
HMP Strange Loops: #1, 148–9, 154; #2, 151, 153–4; #3, 156; *see* Meschonnic
Hoaxes 179–82, 187
Hofstadter, Carol (CHP) 23, 44, 49, 52, 60, 69, 139, 144, 169, 184–6, 195n5, 197n4
Hofstadter, Douglas (DHP) 21, 26, 28, 54, 56–7, 71, 86–7, 98, 117, 140, 148, 179, 192n1, 196n6; aggrandizing the "I" 4, 9–10, 22–3, 36, 43, 49–52; on analogy 55–6, 69–70, 110, 197n5; attacking the phenomenology of feeling 36, 52, 76, 86, 137–9, 144–5, 156, 172; defining strange loops 5–6, 79, 84, 87, 89, 91, 99–100, 127, 135, 141–2, 175; denying external access to the interiority of the "I" 44, 48, 52, 72, 100, 183; on hoaxes and illusions 179–81, 191; on indexical self-reference 2, 5, 11, 38–9, 93, 97; on literal translation as mindless 22–3, 31–9, 53–4, 60; on "locking-in" 26, 76, 131–2, 137, 157, 167, 169, 172; and the multiplicity of strange loops 3–4, 54–5, 75–6, 106, 145, 174, 185; on paradoxes of translation 6–8, 38–9; on partial marital fusion 23, 44, 52, 60, 69, 139, 184–5, 197n4; and physical reductivism 27, 53, 181, 191; on sense-for-sense translation as the non-strange-loopy "solution" 22, 39, 44, 175; not theorizing the

strange loops of translation 1–2, 5, 10–1, 13, 15, 18–9, 22, 38–9; tracking the emergence of the "Ontological I" 91–4, 172–3, 183, 190; translating *Onegin* 193-6n5
Holberg, Ludvig (LHP) 188
Hölderlin, Friedrich (FHP) 125, 148, 199n4
Homer 16, 188
Homolingual address, regime of (Sakai) 128–9, 135
House, Juliane 35
Humboldt, Alexander von 57

I Am a Strange Loop (Hofstadter) 2, 4–5, 10, 21–3, 38, 49–50, 70, 72, 100, 144–5, 184
Icosis (Robinson) 71, 75–6, 128, 143–4, 156–8, 169, 200n3
Idealism, Kantian 165
Identity-formation 71, 98, 127
Illocutionary force (Austin) 61, 101, 105
Illusions 179–82
Impola, Richard 19–20
"Impossible Effigies" (Tiffany) 185
Incoherently written source text 13–6, 43–4, 96
Indexicality 2, 5, 11, 38–9; and shifters 92–3
Interiority of the "I" 43–4, 52; in the source author 13–4, 27, 51, 183, 185, 188, 200n1
Interpretant (Peirce) 142–3
"Interpretranslation" (Tekwa) 34
Intersubjectivity, of rhythm 26, 149, 151–2
Intervenience, in translators 163
Intralingual translation (Jakobson) 99
Introducing Translation Studies (Munday) 196n2
Испытание иностранным/*Ispytanie inostrannym* (Hölderlin/Berman, Avtonomova) 121, 125 *see also Erfahrung des fremden, L'épreuve de l'étranger*
"Iterative scheme" (Eco) 16–7

Jaitner, Sabina Folnović (SFJP) 25, 110–1
Jakobson, Roman 99; interlingual/intersemiotic translation 1

James, William 141
Jerome (ESHP) 22, 39–43, 53, 196n2

"Kääntämisen kääntöpiirit/The Tropics of Translation" (Robinson) 140–1
Kankimäki, Mia (MKP) 45, 139
Kant, Immanuel 168; post-Kantian 165, 170; pre-Kantian 167
Kilpi, Volter 193n4
Kim Hyesoon (KHP) 188, 190
Kitsch 179, 182, 185–6
Kivi, Aleksis (AKP) 19–21, 188
Knowledge 54–5, 64; -exchange 24, 108, 146; as omniscience (Díaz-Diocaretz) 76–7; rhythmic 149; -transfer, as knowledge-translation 146, 148, 151–4, 156–8
Kohut, Heinz 170, 172
Kotzebue, August von 57
Kousbroek, Rudy 38–9
Kristeva, Julia 146
"Kugelmass, Translator" (Robinson) 37

L'épreuve de l'étranger (Hölderlin/Berman) 125, 199n4; *see Erfahrung des fremden, die*
Lacan, Jacques (JLP) 74, 182, 198n2
Lacoue-Labarthe, Philippe 146, 149
Ladmiral, Jean-René 119
Latour, Bruno 161
Leaves of Grass (Whitman) 145
Lefevere, André 97
Level-shifts, analogical 58, 69–70, 93, 99
Lexical simplification 19–20, 32
Liar paradox 8, 24, 27–8, 101–2
"Liar Paradox Monism" (Robinson) 27
Literal translation 23, 31–2, 35, 37, 39, 46, 98, 111, 114, 194-5n5, 196n1; stylized 38, 53, 59, 63, 124–5, 127
Liu, Lydia H. (LHLP) 26, 109, 127–8, 130–1, 133, 135
Locking-in (Hofstadter) 26, 75–6, 167, 169, 172; fueled by affect 76, 137–8, 156–7, 175; loop (Hofstadter/Liu) 131–2, 135
"Logic of prohibition" (Göransson) 184–5
Logos (Aristotle) 104

Lotbinière-Harwood, Susanne de 163
Luther, Martin (MLP) 11–3, 188–9

Machine translation 14, 43, 98; post-editing of 34–5
Marais, Kobus (KMP) 26, 84, 158–70, 176
Marx, Karl 133
Matson, Alex 19–20
McSweeney, Joyelle (JMP) 186
"Meaning of Rhythm, The" (Aviram) 146
"Meat-infatuated philosophers" (Hofstadter) 138, 144
Melnick, David 196n1
Men in Aïda (Melnick) 196n1
Mendonça Cardozo, Mauricio 24, 107–9
Mensah, Patrick (PMP) 102–5
Merleau-Ponty, Maurice (MMPP) 155–6
Meschonnic, Henri (HMP) 26, 145–58, 200n2
Mesku, Melissa (MMP) 4–5, 28
Midsummer Night's Dream, A (Shakespeare) 66, 197n5
Mikra (reading-aloud) 153
Mille plateaux (Deleuze and Guattari D&GP) 187
Minoritization (Deleuze and Guattari D&GP) 188, 200n1
Mirror neurons 156, 169
Mommy Must Be a Mountain of Feathers (Kim/Choi) 189
Monty Python and the Holy Grail 8
Morton, Timothy (TMP) 182, 186
Mother-child dyad 171–3, 176
Mouthability of translation (Meschonnic) 150–5, 157
Movement, entelechial 155
Munday, Jeremy 163, 196n2
Mutual gaze interactions (Schore) 171
My Silver Planet (Tiffany) 184
"Myth of Superman, The" (Eco) 16

Nabokov, Vladimir (VNP) 193-6n5
Napoleon Bonaparte 57
Narratoriality, of the translator 49, 163
Necropastoral (McSweeney) 186
Negation (Freud) 133
Neuroscience: cognitive and affective 141; social 169; socioaffective 27

Nida, Eugene A. (EANP) 54
Niranjana, Tejaswini 97
"Noble rust" (Schlegel) 17
Nonlinear systems theory 161
Norms: of sexuality 72; of translation 13–4, 16, 19, 32, 51, 176
Nummisuutarit (Kivi) 188

Objectivism 10, 167–8, 170
Odyssey (Homer) 16–7
Omniscient Reader (Díaz-Diocaretz) 72, 76–9, 81–2
On Christian Doctrine (Augustine) 40
"On the Different Methods of Translating" (Schleiermacher) 23, 53–65
One (pronoun) 103–8, 110–2, 114–6, 122, 127, 133–4; *see also das Man*, We
"One Land, One People, One Language" 64; *see Ein Land, Ein Volk, Eine Sprache*
Ontological I (Hofstadter) 91
Orality of translation (Meschonnic) 151–5

Pale Fire (Nabokov) 193n4
Pandemonium (Dennett) 3; in analogies 2, 6, 15, 55, 169, 172, 175, 188; pronouns of 2–5; sharing of demons in 13, 23, 44–53, 139–40
Panicked heterosexuality (Butler) 182
Paradise Lost (Milton) 3
Paradoxes of translation 2, 5–11, 13, 18–9, 22, 39
Paranoid reading (Sedgwick) 117, 160
Pathos (Aristotle) 124; convergence with *ēthos* 104, 123
Peirce, Charles Sanders (CSPP) 108; on endless semiosis 90, 197n2; and triads 27–8, 142–4
Peirce's Theory of Signs (Short) 197n2
Perelman, Chaim (CPP) 153
Performative (Austin) 24, 105, 179; and affect 123–4; as embodied performance 18, 108, 150, 152–3, 155; as an I-you dyad 25, 101, 110, 128; and national identity 129; in translating/interpreting 105–7; *see also* Sedgwick, Periperformative
Performative Linguistics (Robinson) 176

Periperformative (Sedgwick) 18, 107; and the aporia 102–4; and audience-effects 21; and the conference interpreter 106; and crowd-sourced TQA 25, 120, 122–4, 126; and *das Man* (Heidegger) 127, 198n1; and disciplinarity 109; and "one" 104, 111, 134; and national identity-formation 127–9; and the reader of philosophical texts 110; and resistance 179; and translation norms 179; and (un-)translatability 25, 116–7, 119; *see also* Performative
Perlocutionary effect (Austin) 61, 101, 105
Peters, John Durham 183–4
Phenomenology 53, 70, 86, 108; of feeling 36–7, 59; of language learners (FSP) 54; and "locking in" 151; of marriage 49; of participation in rhythm 150; of periperformative witnessing 106; of reading 151; of relearning 200n4; of sense-for-sense translation 49; shared 52; of theatricality 6, 180, 183; of translational enmity (FSP) 61–2
Phenomenology of Perception (Merleau-Ponty) 155–6
Philosophy's Treason (Spitzer) 24, 105, 110, 116
Pistis (Aristotle) 27, 124; and *ēthos* 123
"Pivotal Status of the Translator's Habitus, The" (Simeoni) 176
Plato (PP) 67, 146–8
Powers, Richard 66–8
Principia Mathematica (Whitehead and Russell) 1, 137, 184
Pseudotranslation (Toury) 15, 67
Psychosomatic Medicine and Contemporary Psychoanalysis (Taylor) 172
Pushkin, Aleksandr Sergeyevich 2, 9, 193-6n5
Pym, Anthony 199n3; on lexical simplification 19, 32; on translational performatives 105–6

Radio Corpse (Tiffany) 184
Re-belle et Infidèle (Lotbinière-Harwood) 163

Reductivism: in the DHP 31, 40, 139, 181, 192n1; in the KMP 170
"Relational reason" (Mendonça Cardozo) 108
Rhetoric 9, 47, 51, 103–4, 160, 182; and aporias 24, 102, 105; and the audience-effect 18, 69; and the constitution of the "I" 48, 54, 98; and constructs 6, 10, 16, 44, 75–6; and rhythm 75
Rhetoric (Aristotle) 192n2
Rhythm 20, 45–6, 75, 174; semantic history of (Aviram) 146–8; mouth-and-ear for (Meschonnic) 157; serial and intersubjective (Meschonnic) 26, 145, 148–9, 152–8
"Rhythm as Knowledge-Translation, Knowledge as Rhythm-Translation" (Robinson) 200n2
Rich, Adrienne (ARP) 72–3, 76–8, 81, 83
Richards, I.A. 185
Ricoeur, Paul 118
"Rigidism" (Hofstadter) 39, 53
Russell, Bertrand 137, 184

Saarikoski, Pentti 69
Sagan, Françoise (FSaP) 2, 9–10, 22, 40, 49–52, 70
Sakai Naoki (SNP) 8, 26, 128–30, 132–5, 146, 199n5
Salinger, J.D. 69–70
Schäffner, Christina (CSP) 71
Schiavi, Giuliana 49, 107
Schlegel, A.W. 17
Schleiermacher, Friedrich (FSP) 23, 53–65, 83, 116, 124, 126, 199n4
Schleiermacher's Icoses (Robinson) 27, 56, 200n3
Schore, Allan (ASP) 170–3
Schwarze Hefte (Heidegger) 112
Schweetz, Vanellope von (*Wreck-it Ralph*) 135
Scientism 167
Sedgwick, Eve (ESP): on paranoid reading 117, 160; on the periperformative 18, 24–5, 101, 104, 106–7, 110–1, 123, 134

Sein und Zeit (Heidegger) 67, 110
Seitsemän veljestä (Kivi) 19; see Brothers Seven
Self psychology (Kohut) 170
Self-reference 2, 7, 11, 17–8, 44, 93, 97–8; *see also* Indexicality, Shifters
Self-Representational Approaches to Consciousness (Williford and Kriegel) 2
Selfobject (Kohut) 172, 176
Semiotranslating Peirce (Robinson) 27, 200n3
"Sendbrief vom Dolmetschen" (Luther) 11–13; *see* "Circular Letter on Translation"
Sense-for-sense translation 19, 22–3, 34, 36–45, 48–9, 54, 179, 184
Septuagint 40–2, 196n3
Shakespeare, William (WSP) 6, 20–1, 66, 188, 197n5; author-function of 93–4
Sharing: demons (Dennett) 23, 139–40; qualia (Peirce/Robinson) 44; consciousness (Hofstadter) 184–5
Shifters 92–3; *see also* Indexicality, Self-reference
Shklovsky, Viktor 56
Short, T.L. 197n2
Silverman, Sarah 135
Simeoni, Daniel 176
"Simmballs" (Hofstadter) 137, 139
Simon, Sherry 199n2
Simulation 35–6, 69; by the mirror neurons 139, 169; of the source reader for the target reader 54, 59, 62, 125–6
Sinthome (Lacan/Morton) 182
"Sixteen Avant-Garde Perspectives on World Literature and the Translator's (In)visibility" (Robinson) 200n1
Snellman, J.V. (JVSP) 188
Social constructivism (Vygotsky) 26, 80, 170, 172; attacked by Marais 158–60, 162, 165–6, 167, 170; of translation 174–6
Socioaffective 123, 175, 199n4; ecologies 38, 124, 144; neuroscience 27, 108; periperformativity (Sedgwick) 106–7, 126

Somatic (Robinson) 76, 123, 139, 200n3; in Capgras 66, 68; markers (Damasio) 169–71, 173, 175
Somatic-marker hypothesis (Damasio) 170
Somatomimesis 169
Sovereign Ego (Hofstadter) 4
Spatial metaphors 115–6
Speaking into the Air (Peters) 183
Spitzer, D.M. 104–5, 107, 110
Steiner, George (GSP) 17, 112, 190, 199n2
Stolze, Radegundis 27
Strange economy (Tiffany/Göransson) 185
Strange loops, defined 5–6, 79, 84, 87, 89, 91, 99–100, 127, 135, 141–2, 175; and pandemonial pronouns 2–5; and queer/transgender 3, 186; theorized in *Gödel, Escher, Bach* (Hofstadter) 1; theorized in *I Am a Strange Loop* (Hofstadter) 2, 5; of translation, *passim*; used by other scholars 1; *see also* Hofstadter, Douglas
Subject (Lacan) 74–5
Subject in transit (Sakai) 133, 135
Subjectivity 21; and *Dasein* (Heidegger) 110; -formation 22; homolingual (Sakai) 128–9; and rhythm (Aviram/Meschonnic) 147–9, 153–5; transitional (Sakai) 133–5; in the translator-function (Díaz-Diocaretz) 75
Supplement, logic of the (Derrida) 133

Te'amim (Masoretes/Meschonnic) 26, 150
Tekwa, Kizito (KTP) 33–6
That Mad Ache (Sagan/Hofstadter) 36, 40, 49
Theatricality 6, 180; avant-garde 183
Thousand Plateaus, A (Deleuze and Guattari D&GP/Massumi) 187
Tiffany, Daniel (DTP) 182, 184–5
Ton Beau de Marot, Le (Hofstadter) 1–2, 5, 21, 28–9, 38, 100, 193n5, 196n6
Touching Feeling (Sedgwick) 24, 101, 106, 117, 134, 160
Toury, Gideon (GTP) 15, 19, 32, 71, 96
Traduttore, traditore 8–9

Transgender, Translation, Translingual Address (Robinson) 3, 135
Transgressive Circulation (Göransson) 27–8, 179, 191
Transgressive circulations (Peters) 27, 183, 185
Translatability, global 131; *see also* (Un)translatability
Translating Poetic Discourse (Díaz-Diocaretz) 70
Translating the Monster (Robinson) 193n4
Translation and Conflict (Baker) 158
"Translation and Normativity" (Hermans) 71
Translation and Rewriting in the Age of Post-Translation Studies (Gentzler) 66
Translation and Subjectivity (Sakai) 8, 199n5
Translation and the Problem of Sway (Robinson) 99, 200n1
Translation as Intervention (Munday) 163
Translation Studies 22, 24, 78, 128; post- 66; primal scene of 130; *see also* Descriptive Translation Studies, Critical Translation Studies
Translation: archaizing 16–21; assessment as "bad" 121–4, as "excellent" 124–6; as co-creation (Hofstadter) 196n4; domesticating 63–4, 200n1; as experienced by the body-becoming-mind 148, 150–2; foreignizing 53, 59, 61–3, 116, 124, 126, 200n1; homophonic 38; interlingual (Jakobson) 1; intersemiotic (Jakobson) 1; as intersubjective 26, 148–9, 151–2; intralingual (Jakobson) 99; literal 23, 31–2, 35, 37, 39, 46, 53, 59–62, 98, 111, 125, 194–5n5, 196n1; machine 14, 34–6, 43, 98; modernizing 16; as mouthable/oralizable (Meschonnic) 150–5; norms of 13–4, 16, 19, 32, 51, 115, 176, 198n1; paradoxes of 2, 5–11, 13, 18–9, 22, 39; as repetition (Sakai) 128; as rhythmic 148–9, 154; representation of (Sakai) 128, 135; as serial 148–50; sense-for-sense 19, 22–3, 34, 36–45, 48–9, 54, 179, 184; universals of 19–20, 32; violence of (Steiner) 112–3; word-for-word 22–3, 31, 33, 36–7, 40–3, 53–4, 60, 179, 193n5
Translationality (Robinson) 56, 66
Translator-function (Díaz-Diocaretz) 24, 42, 70, 71–84, 86, 115, 170, 176, 179; in Hofstadter 12; in Robinson 12, 71; in Luther 12–3; in Arrojo 71, 84–6, 197n1; in Hermans 71, 94–7; transmajoritizing 188
Translator, as intervenient 163; as the glitch 135; as narrator 49, 163; *see also* Glitch loop, Narratoriality, *Translation as Intervention*
"Translator, Trader" (Hofstadter) 2, 4–9, 21–3, 31, 35, 38, 49, 175, 196n4
Translator's (in)visibility (Venuti) 84–5, 95, 190
Translator's Turn, The (Robinson) 26, 139–45, 155
Translinguality (Robinson) 135
Transmajorization 188, 200n1
Transminoritization 188
"Trial of foreign" (Hölderlin/Berman/Avtonomova/Gukasyan) 121, 125–6; *see also Erfahrung des fremden, L'épreuve de l'étranger,* Испытание иностранным
Tymoczko, Maria (MTP) 96

"Uncertain agency" (Hertz/Sedgwick) 106, 123
(Un)translatability 25, 67, 109–19; global 131

Venuti, Lawrence (LVP) 53, 63, 84, 116, 124, 126, 190, 200n1
"Verbomania" (Richards) 185
Verfremdung (Brecht) 59
Vihelmaa, Ella 199n1
Violence, in translation (Steiner) 112; in interpretation (Heidegger) 112–3
Vocabulaire européen des philosophies (dictionnaire des intraduisibles) (Cassin) 110, 117

Vulgate Bible (Jerome) 41–2
Vygotsky, Lev 165, 170

We (pronoun) 52, 102–4; *see also* One
Wenger, Etienne 146
Western Translation Theory from Herodotus to Nietzsche (Robinson) 11, 17, 39, 40, 53, 62, 124
What I Am (Heidegger) 114, 116, 121; *see* Dasein
"What Is an Author" (Foucault) 24, 72, 87, 91–2, 197n1
What is Translation? (Robinson) 70–1
What One Does (Heidegger) 25, 114–6, 122, 127; *see das Man*

Whitehead, Alfred North 137, 184
Whitman, Walt (WWP) 52, 145
Who Translates? (Robinson) 3, 196n3
Word-for-word translation 22–3, 31, 33, 36–7, 40–3, 53–4, 60, 179, 193n5
Wordsworth, William 182–3
Wreck-it Ralph (Moore) 135
"Writing in No Man's Land" (Bassnett) 199n2
Wrong-Place Paradox (Hofstadter) 7–9, 18
Wrong-Tongue Paradox (Hofstadter) 6–7, 10, 13

Zaum 38, 196n1
Zavala, Iris M. 76

www.ingramcontent.com/pod-product-compliance
Lightning Source LLC
Chambersburg PA
CBHW062217300426
44115CB00012BA/2111